Design and Control of

AUTOMOTIVE PROPULSION SYSTEMS

MECHANICAL and AEROSPACE ENGINEERING

Frank Kreith & Darrell W. Pepper
Series Editors

RECENTLY PUBLISHED TITLES

Design and Control of
AUTOMOTIVE PROPULSION SYSTEMS

ZONGXUAN SUN
GUOMING G. ZHU

CRC Press
Taylor & Francis Group
Boca Raton London New York

CRC Press is an imprint of the
Taylor & Francis Group, an **informa** business

CRC Press
Taylor & Francis Group
6000 Broken Sound Parkway NW, Suite 300
Boca Raton, FL 33487-2742

First issued in paperback 2017

ISBN-13: 978-1-4398-2018-6 (hbk)
ISBN-13: 978-1-138-74866-8 (pbk)

Library of Congress Cataloging-in-Publication Data

Sun, Zongxuan.
   Design and control of automotive propulsion systems / Zongxuan Sun, Guoming (George) Zhu.
      pages cm -- (Mechanical and aerospace engineering)
   Includes bibliographical references and index.
   ISBN 978-1-4398-2018-6 (hardback)
   1. Automobiles--Motors--Design and construction. 2. Automobiles--Power trains--Design and construction. 3. Propulsion systems--Automatic control. I. Zhu, Guoming (Engineer) II. Title.

TL210.S9435 2014
629.25--dc23   2014026569

Visit the Taylor & Francis Web site at
http://www.taylorandfrancis.com

and the CRC Press Web site at
http://www.crcpress.com

Contents

Preface

Transportation consumes about 30% of the total energy in the United States. In many emerging markets around the world, transportation, especially personal transportation, has been growing at a rapid pace. Consequently, energy consumption and its environmental impact are now among the most challenging problems humans face. From a technical perspective, construction machinery and agriculture equipment share similar challenges, as all mobile applications have to carry energy onboard and convert energy into mechanical motion in real time to meet the demand of the specific function. The objective of this book is to present the design and control of automotive propulsion systems in order to promote innovations in transportation and mobile applications, and therefore reduce their energy consumption and emissions.

There are two unique features of this book. One is that given the multidisciplinary nature of the automotive propulsion system, we adopt a holistic approach to present the subject, especially focusing on the relationship between propulsion system design and its dynamics and electronic control. A critical trend in this area is to have more electronics, including sensors, actuators, and controls, integrated into the powertrain system. This is going to change the traditional mechanical powertrain into a mechatronic powertrain. Such change will have profound impact on the complex dynamics of the powertrain system and create new opportunities for improving system efficiency. The other is that we cover all major propulsion system components, from internal combustion engines to transmissions and hybrid powertrains. Given the trend of vehicle development, system-level optimization over engines, transmissions, and hybrids is necessary for improving propulsion system efficiency and performance. We treat all three major subsystems in the book.

Chapter 1 presents the background of the automotive propulsion system, highlights its challenges and opportunities, and shows the detailed procedures for calculating vehicle power demand and the associated powertrain operating conditions. Chapter 2 presents the design, modeling, and control of the internal combustion engine and its key subsystems: the valve actuation system, the fuel system, and the ignition system. Chapter 3 presents the operating principles of the transmission system, the design of the clutch actuation system, and transmission dynamics and control. Chapter 4 presents the hybrid powertrain, including the hybrid architecture analysis, the hybrid powertrain model, and the energy management strategies. Chapter 5 presents the electronic control unit and its functionalities, the software-in-the-loop and hardware-in-the-loop techniques for developing and validating control systems.

This book is intended for both engineering students and automotive engineers and researchers who are interested in designing the automotive propulsion system, optimizing its dynamic behavior, and control system integration and optimization. For the engineering students, this book can be used as a textbook for a senior technical elective class or a graduate-level class. Similar content has been taught in a graduate-level class at the University of Minnesota and received very positive feedback from students. For automotive engineers, the book can be used to better understand the relationship between powertrain system design and its control integration, which is traditionally divided into two different functional groups in the automotive industry. It will also help automotive engineers to understand advanced control methodologies and their implementation, and facilitate the introduction of new design and control technologies into future automobiles.

We thank and acknowledge our graduate students for their contributions to the research work represented in the book. We especially want to thank Yaoying Wang, Yu Wang, Xingyong Song, and Xiaojian Yang for their help with editing and proofreading of the book.

Zongxuan Sun and Guoming Zhu

About the Authors

Dr. Zongxuan Sun is currently an associate professor of mechanical engineering at the University of Minnesota, Minneapolis. He was a staff researcher from 2006 to 2007 and a senior researcher from 2000 to 2006 at the General Motors Research and Development Center, Warren, Michigan. Dr. Sun received his BS degree in automatic control from Southeast University, Nanjing, China, in 1995, and the MS and PhD degrees in mechanical engineering from the University of Illinois at Urbana-Champaign, in 1998 and 2000, respectively. He has published more than 90 refereed technical papers and received 19 U.S. patents. His current research interests include controls and mechatronics with applications to the automotive propulsion systems. Dr. Sun is a recipient of the George W. Taylor Career Development Award from the College of Science and Engineering, University of Minnesota, the National Science Foundation CAREER Award, the SAE Ralph R. Teetor Educational Award, the Best Paper Award from the 2012 International Conference on Advanced Vehicle Technologies and Integration, the Inventor Milestone Award, the Spark Plug Award, and the Charles L. McCuen Special Achievement Award from GM Research and Development.

Dr. Guoming G. Zhu is a professor of mechanical engineering and electrical/computer engineering at Michigan State University. Prior to joining the ME and ECE departments, he was a technical fellow in advanced powertrain systems at Visteon Corporation. He also worked for Cummins Engine Co. as a technical advisor. Dr. Zhu earned his PhD (1992) in aerospace engineering at Purdue University. His BS and MS degrees (1982 and 1984, respectively) were from Beijing University of Aeronautics and Astronautics in China. His current research interests include closed-loop combustion control, adaptive control, closed-loop system identification, linear parameter varying (LPV) control of automotive systems, hybrid powertrain control and optimization, and thermoelectric generator management systems. Dr. Zhu has more than 30 years of experience related to control theory and applications. He has authored or coauthored more than 140 refereed technical papers and received 40 U.S. patents. He was an associate editor for ASME *Journal of Dynamic Systems, Measurement, and Control* and a member of the editorial board of *International Journal of Powertrain*. Dr. Zhu is a fellow of the Society of Automotive Engineers (SAE) and American Society of Mechanical Engineers (ASME).

1

Introduction of the Automotive Propulsion System

1.1 Background of the Automotive Propulsion System

1.1.1 Historic Perspective

Throughout human history, transportation of people and goods has always been a critical part of society. For a very long time (until the 19th century), this was accomplished by either human- or animal-driven vehicles. The steam engine fundamentally changed the transportation system, mainly by powering boats and trains with some applications for automobiles. At the end of the 19th century, the invention of the internal combustion engine (ICE) led to a complete revolution of both personal and commercial transportation. Over the past 100 years, the ICE has dominated the automotive propulsion system. This is mainly due to the energy density of the liquid fuel and the power density of the ICE. For the first time in human history, the ICE enables the controlled extraction of chemical energy in hydrocarbon fuels into mechanical motion through cyclic exothermic chemical reactions with high power density.

Tremendous improvement has been achieved for optimizing engine performance, efficiency, and emissions. Today's ICE is a much more complex machine than its ancestor of a century ago. New technologies appear in nearly every subsystem of the ICE: air intake and exhaust system, fuel delivery and injection system, ignition system, cooling system, lubrication system, aftertreatment system, materials and manufacturing technology, and sensing and control system. This is the result of century-long efforts of continuous innovations involving science, engineering, and technology. The hallmark of such innovations is their multidisciplinary nature. This involves mechanical engineering, electrical engineering, chemistry and materials, etc. If we zoom into the specific disciplines, they include thermodynamics, fluid mechanics, heat transfer, chemical reaction, design and manufacturing, controls, etc. This multidisciplinary nature has served us well, but it also reveals the difficulties and complexities we will face as the technology evolves going forward.

1.1.2 Current Status and Challenges

As we entered the 21st century, new challenges emerged for the transportation system. On the one hand, it became an integral part of society. Both personal and commercial transportation through on-road vehicles became necessary tools for everyday life and economic activities. Off-road vehicles such as construction machinery and agriculture equipment also experienced significant growth for improving productivity in many industries and farming. On the other hand, the growing number of vehicles around the world poses a serious challenge to the sustainability of transportation and its impact on the environment.

There are about 850 million automobiles in the world today, with a projected number of 2.5 billion by year 2050. These enormous numbers once again bring up a question that was debated more than a hundred years ago: What is the best propulsion system for automobiles, and what are the energy sources that can sustain transportation? To answer such questions, research work for improving the efficiency of the ICE-based propulsion system, designing alternative propulsion systems, and developing renewable energies is being pursued. A good example is the emergence of hybrid vehicles more than 10 years ago. The hybrid powertrain is the first major change from the conventional powertrain by adding alternative power sources such as electrical power or fluid power to the system. More technical innovations are expected in the coming years that could reinvent the automotive propulsion system. To facilitate such innovations, this book is targeted to introducing the design, modeling, and control of the current automotive propulsion system, as well as presenting and discussing future trends.

1.1.3 Future Perspective

There have been numerous predictions and debates on the time when fossil fuel will be exhausted. Likely this is still a subject for debate even today. However, what is clear and less controversial is that global energy consumption has been growing at an unprecedented pace, conventional oil and gas supplies are being depleted, and there are tremendous concerns regarding the environmental impact of greenhouse gas emissions. To account for these challenges, three types of energy sources have been proposed for transportation: liquid and gaseous fuels from both fossil and renewable sources, electricity, and hydrogen. The corresponding powertrain systems are internal combustion engine, electric propulsion, and fuel cell. While there are several discussions on the advantages and disadvantages of different powertrain systems, their fate, to a large extent, will be determined by the competition among the various energy sources.

The main advantages of liquid fuels are the energy density, ease of handling, and transportation. So far, liquid fuels still have a clear advantage (order of magnitude) over any other energy sources for the ability to store energy per unit volume or weight. It is also fairly easy to replenish the fuel once it is consumed. The current practice of pumping gasoline at the gas station is in fact adding several hundred megajoules per minute into the vehicle. The extensive network of gas stations makes fuel transportation and storage convenient and cost-effective. Those seemingly obvious features (energy density, easy to refuel and transport) are indeed the key factors that are needed for transportation energy supply. Using electricity as the fuel for transportation has the advantage of centralized emissions control since the emissions occur at the power plant rather than at the individual vehicle. It is also more versatile to incorporate renewable sources such as wind and solar energy. The main challenge for using electricity for transportation or mobile applications is the battery. To be competitive at large-scale deployment, the battery needs to have energy density that is comparable with the liquid fuel and easy and fast to recharge. The hydrogen fuel uses the most abundant element in the world and produces no emissions at the vehicle level. However, production of the hydrogen fuel, as well as its transportation and storage, still faces many technical challenges. Although there are many studies for comparing the well-to-wheel energy consumption of the different energy sources, this is not the focus of this book. Given the fact that liquid fuel will likely still dominate the energy supply for transportation for the foreseeable future, this book will focus on the ICE-based automotive powertrain system while presenting the alternative powertrain systems where appropriate.

1.2 Main Components of the Automotive Propulsion System

For any mobile applications, the energy source must be carried onboard and converted into mechanical energy and transferred to the wheels in real time (Figure 1.1). Main components of the automotive propulsion system include the engine and the transmission.

The engine is a device that facilitates the combustion process and extracts the chemical energy into thermal energy and further converts it into mechanical work. The combustion is an exothermic process that releases heat through the chemical reaction of two reactants: the fuel and the air (oxygen). The combusted gas with elevated temperature drives the piston that produces mechanical work. To operate in a cyclic fashion, the engine follows the Otto cycle for gasoline engines and Diesel cycle for diesel engines. Analysis has shown there are many irreversible processes during the operation of the ICE that lead to the efficiency degradation of the system. New designs that target these losses, as well as the energy in the exhaust, have been proposed. One objective of this book is to introduce these new designs and analyze their impact on the engine.

The transmission is a device that transfers the mechanical output of the engine to the wheels of the vehicle. In theory, the transmission is not necessary if the engine's torque, power, and efficiency are not functions of speed. So to optimize the torque, power, and efficiency of the engine, a transmission is required to change the operating condition (speed and load) from the vehicle operating condition in real time. The most commonly used transmission mechanisms are gears, which provide different ratios between the vehicle and the engine. To switch between different ratios, actuators are required to change the gears. This can be done through either a human driver (manual transmission) or an electronically controlled system (automatic transmission). The efficiency of the transmission system is determined by the efficiency of the gears and the actuation system.

The fundamental challenge that limits the efficiency of the engine and transmission is the dynamic operating requirement of the vehicle in real time. The typical power demand for an automobile can span a ratio of 10. For example, a vehicle cruising on the highway may only need 10 kW to maintain the required constant speed, while the vehicle could demand 100 kW for wide-open throttle operation during acceleration. Unfortunately, the engine is sized for the most demanding performance criterion, which forces it to operate at part load conditions in many scenarios. As we know, the ICE and transmission efficiency is a function of the operating condition. To improve the system efficiency, we must understand the dynamic power demand of the vehicle, the root cause of the inefficiency, and then propose the corresponding solutions.

1.3 Vehicle Power Demand Analysis

As mentioned before, a lot of challenges associated with the automobile powertrain system are due to the mobile nature of the application. In this section, we first study how

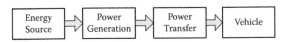

FIGURE 1.1
Main components of the automotive propulsion system.

to calculate the vehicle tractive force, and then use it to analyze the vehicle power demand during various driving operations [1–3].

1.3.1 Calculation of Vehicle Tractive Force

For an automobile in motion, the typical resistance forces include rolling resistance due to tire and road interaction, wind resistance due to air and vehicle interaction, grade resistance due to the various grades of the road, and acceleration resistance due to the need to accelerate the vehicle mass.

As shown in Figure 1.2, the total tractive force for the vehicle is

$$F_T = F_R + F_W + F_G + F_A \tag{1.1}$$

where F_T is the total tractive force at wheels (N), F_R is the rolling resistance force (N), F_W is the wind resistance force (N), F_G is the grade resistance force (N), and F_A is the acceleration resistance force (N).

We first show how to calculate the resistance forces and then use them to calculate the vehicle performance limit.

The rolling resistance force is

$$F_R = K_R \cdot W \cdot \cos(\theta) \tag{1.2}$$

where K_R is the rolling resistance coefficient (for typical values, see Table 1.1), W is the vehicle weight (N), and Θ is the road grade angle (radian).

The wind resistance force is

$$F_W = K_W \cdot A \cdot V^2 \tag{1.3}$$

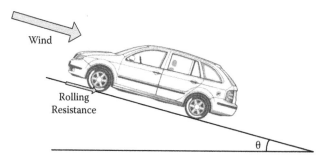

FIGURE 1.2
Vehicle resistance forces.

TABLE 1.1

Typical Values for the Rolling Resistance Coefficient

K_R	Road Condition
0.01	Good paved roads
0.015	Average paved roads
0.02~0.025	Good gravel or soil
0.1~0.15	Sand

TABLE 1.2

Typical Values for Wind Resistance Coefficient

K_W	Vehicle Type
0.2995	Passenger cars
0.4313	Small trucks
0.599	Large trucks

where K_W is the wind resistance coefficient $(N/(m^2/s)^2)$, A is the vehicle frontal area (m^2), and V is the vehicle speed (m/s) (Table 1.2).

The grade resistance force is

$$F_G = W \cdot \sin(\theta) \tag{1.4}$$

where W and θ are the same as defined before.

The acceleration resistance force is

$$F_A = \left(\frac{W}{g}\right) a = W\left(\frac{a}{g}\right) \tag{1.5}$$

where W is the same as defined before, a is the vehicle acceleration (m/s^2), and g is the gravitational constant (9.8 N/kg).

When calculating the acceleration resistance force, if necessary, the equivalent vehicle weight that includes the effective weight of the powertrain rotating components can be used. This is because during vehicle acceleration, not only is the vehicle mass undergoing linear acceleration, but the rotational components of the powertrain system (engine inertia, transmission, drive shaft, and tire) are also undergoing angular acceleration, which is directly related to the linear acceleration of the vehicle. Depending on the inertia of the powertrain system relative to the vehicle mass, the equivalent vehicle weight could be significantly higher than the actual vehicle weight.

Now we are going to use the above equations to calculate the vehicle performance limit. What road grade will produce the same resistance force required for a given vehicle acceleration?

Let $F_G = F_A$; we have

$$W \cdot \sin(\theta) = \frac{W \cdot a}{g}$$

So

$$\sin(\theta) = \frac{a}{g}$$

Using this equation, we can calculate the relationship between road grade angle and the acceleration, as shown in Table 1.3.

TABLE 1.3

Vehicle Acceleration and Equivalent Road
Grade Angle

Acceleration	Equivalent Grade Angle (degree)
0	0
$0.1\,g$	5.74
$0.2\,g$	11.54
$0.3\,g$	17.46
$0.4\,g$	23.58

1.3.1.1 Traction Limit

The vehicle tractive force limit is based on the maximum traction available between tires
and the road surface:

$$F_{T-Max} = \mu \cdot W \cdot X \cdot \cos(\theta) \tag{1.6}$$

where W and θ are the same as defined before, μ is the friction coefficient, and X is the
percentage of vehicle weight on driving wheels (40%~60% for 2WD and 100% for 4WD).

1.3.1.2 Maximum Acceleration Limit

The maximum vehicle acceleration is determined by applying the maximum traction force
to accelerate the vehicle without any grade resistance and wind resistance (vehicle speed
at zero):

$$F_{T-Max} = F_R + F_A \Rightarrow \mu \cdot W \cdot X = K_R \cdot W + W \cdot \frac{a}{g}$$

Assume K_R is very small and the vehicle is 4WD; we have

$$\mu = \frac{a}{g}$$

So

$$a_{Max} = \mu g \tag{1.7}$$

1.3.1.3 Maximum Grade Limit

The maximum grade limit is determined by applying the maximum traction force to climb
the grade without any acceleration and wind resistances (vehicle speed at zero):

$$F_T = F_R + F_G \Rightarrow \mu \cdot W \cdot X \cdot \cos(\theta) = K_R \cdot W \cdot \cos(\theta) + W \cdot \sin(\theta)$$

Assume K_R is very small and the vehicle is 4WD; we have

$$\mu\cos(\theta) = \sin(\theta) \Rightarrow \tan(\theta) = \mu$$

So

$$\mu = 1.0 \Rightarrow \theta_{Max} = 45° \tag{1.8}$$

1.3.1.4 Vehicle Power Demand

The required vehicle power at any time instant is the product of the tractive force multiplying the vehicle speed:

$$P = F_T \cdot V \tag{1.9}$$

Example 1.1

Consider a 1500 kg vehicle, 2.5 m² frontal area, $r_{tire} = 0.3$ m, rolling resistance coefficient $K_R = 0.015$, wind resistance coefficient $K_W = 0.3$ N/(m²/s)².

1. Calculate the tractive force required to accelerate at 0.2 g at 70 km/h on a level road.
2. Calculate the power on a level road, steady speed of 90 km/h.
3. Calculate the power required to climb a 5.71° grade at 90 km/h.

Solution

1. Based on Equation (1.1), we have

$$F_T = F_R + F_W + F_G + F_A$$

Since the vehicle operates on a level road ($F_G = 0$),

$$F_T = K_R W + K_W V^2 A + W(a/g)$$

$$= (0.015)1500(9.8) + (0.3)(19.44)^2(2.5) + 1500(9.8)(0.2) = 9100.5 \text{ N}$$

2. Based on Equation (1.9), the vehicle power demand

$$P = (F_R + F_W + F_G + F_A)V$$

Since the vehicle operates on a level road with no acceleration ($F_G = F_A = 0$),

$$P = (K_R W + K_W V^2 A)V$$

$$= [(0.015)(1500)(9.8) + (0.3)25^2(2.5)]25$$

$$= 17.23 \text{ kW}$$

3. Again using Equation (1.9), we have

$$P = (F_R + F_W + F_G + F_A)V$$

Since the vehicle is operating at constant speed ($F_A = 0$),

$$P = (K_R W\cos(\Theta) + K_W V^2 A + W\sin(\theta))V$$

$$= [(0.015)(1500)(9.8)(0.995) + (0.3)25^2(2.5) + 1500(9.8)(0.0995)]25$$

$$= 53.8 \text{ kW}$$

1.3.1.5 Vehicle Performance Envelope

The vehicle performance envelope can be defined by the maximum tractive force, the maximum engine power, and the vehicle resistance as a function of vehicle speed. This envelope illustrates the ideal possible operating range of the vehicle. In actual vehicle operation, the maximum power of the engine may not be accessible at every vehicle speed due to the discrete gear ratios. So the actual operating range may be smaller (Figure 1.3).

1.3.1.6 Vehicle Power Envelope

The vehicle power demand is defined by the maximum engine power, the vehicle resistance as a function of vehicle speed and acceleration or road grade. The power envelope shows clearly the effect of vehicle acceleration on the required power. The rolling resistance and wind resistance always exist during vehicle operation, but the acceleration could add significant power demand to the overall vehicle power (Figure 1.4).

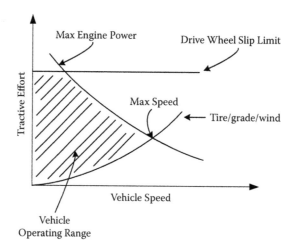

FIGURE 1.3
Vehicle performance envelope.

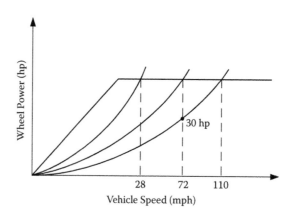

FIGURE 1.4
Vehicle power envelope.

1.3.2 Vehicle Power Demand during Driving Cycles

Based on the vehicle tractive force calculation, we can calculate the required power for propelling the vehicle for different driving cycles [4, 5]. The dynamic behavior in terms of both speed and power of the automobile sets the challenge for the powertrain system.

Example 1.2

Calculate the tractive effort and power demand for a given vehicle during the Federal Test Procedure (FTP) cycle (Table 1.4).

According to Equation (1.1), $F_T = K_R W + K_W V^2 A + W(a/g)$, once tractive force has been calculated, the operating point of the engine (speed ω_e and torque/load T_e) is calculated at each time step by using the following relationship:

$$\omega_e = \frac{r_f r_t}{r_r} V, \quad T_e = \frac{r_r}{r_f r_t} F_T$$

As the FTP cycle begins at a vehicle speed of zero, the vehicle can be assumed to start in first gear at the first time step. These values are then fed into an engine map to determine the required percent throttle to produce the required engine torque at the required engine speed. This percent throttle, along with the vehicle speed, is fed into the transmission shift schedule for a typical automatic transmission with four forward speeds, which determines the shift command (upshift, downshift, or stay in current gear). Note that the percent throttle is calculated only for the purpose of determining gear shifts. This process is repeated at each time step, using the gear commanded in the previous time step to determine engine speed and torque, as well as the shift command for the current time step (which will be used in the next time step).

We can generate the shifting schedule and engine torque/load table for the FTP cycle by using the method described above and use them as inputs to run a driveline dynamic model simulation. Figures 1.5 and 1.6 show the engine map and transmission shift schedule used in this example. Figure 1.7 shows the FTP driving cycle, resulting gear ratio, tractive force, and vehicle power demand during the driving cycle. It is clear that the power demand varies significantly during the driving cycle, and this again sets the fundamental challenge for the powertrain system of mobile applications.

TABLE 1.4

Driveline and Vehicle Parameters

Parameters	Value	Description
Mv (kg)	1400	Mass of the vehicle
K_R	0.015	Rolling resistance coefficient
K_W (N/(m²/s)²)	0.3	Wind resistance coefficient
A (m²)	2.35	Frontal area of vehicle
g (m/s²)	9.8	Acceleration due to gravity
r_f	3.37	Final-drive ratio
r_t	[2.393 1.450 1.000 0.677]	4-speed transmission ratios
r_r (m)	0.325	Rolling radius of the wheel

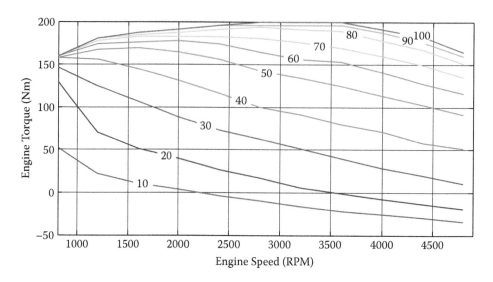

FIGURE 1.5
Engine map showing the contour of throttle as functions of engine speed and torque.

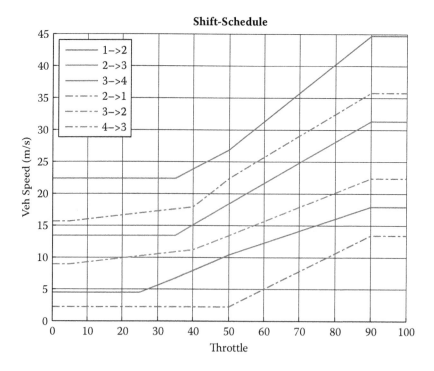

FIGURE 1.6
Transmission shift schedule.

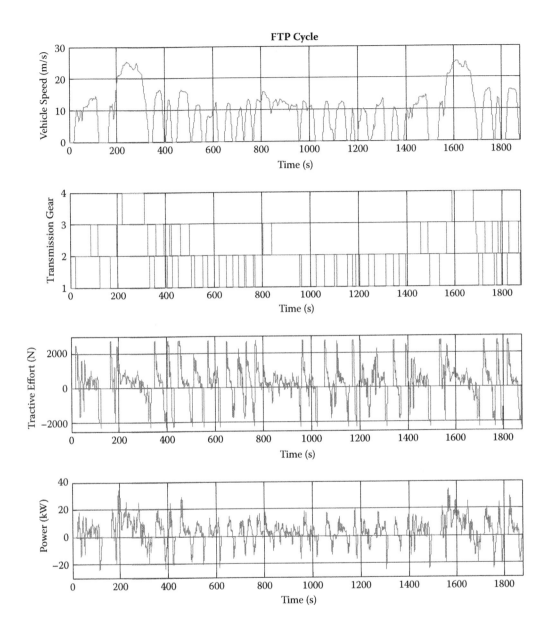

FIGURE 1.7
The FTP driving cycle, the gear ratio, the tractive effort, and the vehicle power demand.

References

1. E. Burke, L.H. Nagler, E.C. Campbell, L.C. Lundstron, W.E. Zierer, H.L. Welch, T.D. Kosier, and W.A. McConnell, Where Does All the Power Go? SAE Technical Paper 570058, 1957.
2. P.N. Blumberg, Powertrain Simulation: A Tool for the Design and Evaluation of Engine Control Strategies in Vehicles, SAE Technical Paper 760158, 1976.
3. R.A. Bechtold, Ingredients of Fuel Economy, SAE Technical Paper 790928, 1979.

4. V. Mallela, Design, Modeling and Control of a Novel Architecture for Automatic Transmission Systems, Master of Science thesis, University of Minnesota, Twin Cities, May 2013.

5. A. Heinzen, P. Gillella, and Z. Sun, Iterative Learning Control of a Fully Flexible Valve Actuation System for Non-Throttled Engine Load Control, *Control Engineering Practice*, 19(12): 1490–1505, 2011.

2

Design, Modeling, and Control of Internal Combustion Engine

2.1 Introduction to Engine Subsystems

The engine subsystems can be divided into the fuel system, ignition system, valve system, exhaust gas recirculation system, turbo-compressor system, etc. This chapter mainly discusses the design, modeling, and control of these subsystems.

For control strategy development, zero-dimensional mean value engine models are widely used [1, 2], due to their simplicity and low simulation throughput. For the engine air handling system or crankshaft dynamics, mean value models are accurate enough since the piston reciprocating movement has less impact on these subsystems than on the combustion process. Therefore, the mean value modeling approach is often used for these subsystems. The disadvantage of mean value engine modeling is that it does not provide detailed information about the engine combustion process, such as in-cylinder gas pressure, temperature, and ionization signals, which have been widely used for closed-loop combustion control [3–5]. The in-cylinder pressure rise is also a key indicator for detecting engine knock [6].

In order to explore the details about the engine combustion process, multizone, three-dimensional computational fluid dynamics (CFD) models with detailed chemical kinetics are presented in [7–9] that describe the thermodynamic, fluid flow, heat transfer, and pollutant formation phenomena of the homogeneous charge compression ignition (HCCI) combustion. Similar combustion models have also been implemented into commercial codes such as GT-Power [10] and Wave. However, these high-fidelity models cannot be directly used for control strategy development since they are too complicated to be used for real-time simulations, but they can be used as reference models for developing simplified (or control-oriented) combustion models for control development and validation purposes.

For real-time hardware-in-the-loop (HIL) simulations, it is necessary to develop a type of combustion model with its complexity in between the time-based mean value models and the CFD models. This motivates the combustion modeling work presented in this chapter. The zero-dimensional (0-D) crank-based combustion model is described in this chapter. Table 2.1 compares the capability of this modeling method with the other two.

The engine valve actuation subsystem employs a camshaft to open and close poppet-type intake and exhaust valves. The camshaft is connected to the crankshaft mechanically to ensure synchronized motion between the intake and exhaust valves and the piston motion. To improve the flexibility of the valve actuation system, a variable valve timing system, variable valve lift, and duration system have been designed to enhance the existing camshaft-based system. Camless systems have also been proposed to replace the camshaft

TABLE 2.1

Features for Different Combustion Models

	0-D Time-Based Combustion Model	0-D Crank-Based Combustion Model	1-D and 3-D CFD Combustion Model
Implementation tool	MATLAB®/ Simulink®	MATLAB®/Simulink® and HIL simulator	GT-Power, Wave, and Fluent
Time cost per cycle	Microseconds	Real time	Minutes to hours
One-zone	Yes	Yes	Yes
Two-zone	No	Yes	Yes
Multizone	No	No	Yes
Chemical concentration	No	No	Yes
Gas-fuel mixing model	No	No	Yes
In-cylinder flow dynamics	No	No	Yes
IMEP	Yes	Yes	Yes
In-cylinder pressure	No	Yes	Yes
In-cylinder temperature	No	Yes	Yes
Ionization signal	No	Yes	No

with an electronically controlled actuator. This chapter will present the design, modeling, and control of the valve actuation system, with a focus on the camless system.

The engine fuel subsystem can be divided into port fuel injection (PFI) and direct injection (DI) systems, where PFI wall-wetting dynamics is the key for accurate fueling control to reach the desired air-to-fuel ratio and engine output torque, while the in-cylinder spray and mixing are also very important for the DI fuel system. This chapter addresses the two key issues for both PFI and DI fuel systems.

Closed-loop combustion control can be used to optimize the combustion process for internal combustion engines. This chapter presents a closed-loop ignition timing control methodology to optimize the engine ignition timing for the best thermal efficiency possible when the engine operates within its knock and combustion stability region.

2.2 Mean Value Engine Model

The control-oriented engine model is mainly used to develop and validate the model-based engine control strategies. Therefore, the model needs to be simple but contains the required engine dynamics. In order to make the HIL simulation possible, the control-oriented engine model shall be able to be executed in real time. Normally, the engine model can be divided into three portions: mean value gas flow model, crank-based combustion model, and cycle-by-cycle event-based model. The following subsections describe the three submodel sets individually.

2.2.1 Mean Value Gas Flow Model

This subsection describes mathematical engine subsystem models whose averaged dynamic behaviors are required for control strategy development and validation, even though they are functions of the engine reciprocating phenomenon. All parameters and variables used in these models are functions of time t.

2.2.1.1 Valve Dynamic Model

The engine valve model, described below, can be used for the intake throttle, the high-pressure (HP) and low-pressure (LP) waste gate, and the exhaust gas recirculation (EGR) valve since these actuators share the same physical characteristic. Assuming that the spatial effects of the connecting pipes before and after these valves are neglected and their thermodynamic characteristics are isentropic expansion [11], the governing equation of the valve model is

$$m_v = C_d A_v \frac{P_{up}}{\sqrt{RT_{up}}} \psi\left(\frac{P_{down}}{P_{up}}\right) \tag{2.1}$$

where

$$\psi(x) = \begin{cases} \sqrt{2x(1-x)}, & \text{if } \frac{1}{2} < x < 1 \\ \frac{1}{\sqrt{2}}, & \text{if } x < \frac{1}{2} \end{cases} \tag{2.2}$$

C_d is the valve discharge coefficient, A_v is the valve open area, P_{up} and T_{up} are the valve upstream pressure and temperature, P_{down} is the valve downstream pressure, and m_v is the mass flow rate across the valve. Note that both C_d and A_v are functions of the valve opening angle θ_v.

2.2.1.2 Manifold Filling Dynamic Model

This subsystem model is mainly used for the intake and exhaust manifolds, intercompressor and interturbine pipes. The receiving behavior is assumed to be an adiabatic process in these manifolds [11]. Their thermodynamic states are uniform over the entire manifold volume, and the manifold temperature is averaged over one engine cycle for this mean value model. Then the governing equation for the manifold filling dynamics is

$$\frac{dP_m}{dt} = \frac{RT_m}{V_m}(m_{in} - m_{out}) \tag{2.3}$$

where P_m is the manifold pressure, T_m is the manifold temperature, V_m represents the manifold volume, and m_{in} and m_{out} are the inlet and outlet air mass flow rates, respectively.

2.2.1.3 Turbine and Compressor Models

The turbocharger can be modeled using the so-called energy conservative equations based upon its steady-state compressor and turbine maps, which can be found in [12, 13]. Notice that the turbo mass flow rate (MFR) and shaft speed in the turbo mapping equations given below are in the so-called reduced form, to make the turbo maps applicable for all inlet conditions. Without this conversion, different turbo maps for each combination of inlet pressure and temperature [14] would be required.

Both turbine and compressor dynamics are described below in Equations (2.4) to (2.11), where P_{in} and T_{in} are either turbine or compressor inlet pressure and temperature,

P_{out} and T_{out} are either turbine or compressor outlet pressure and temperature, N_{turbo} is the turbocharger shaft speed in rpm, and η denotes thermal efficiency.

1. **Turbine mapping:** Turbine maps, f_{turb} and f'_{turb}, in Equations (2.4) and (2.5) are used to calculate the reduced MFR m_{turb} and thermal efficiency η_{turb} based on pressure ratio across the turbine and the reduced turbo shaft speed. The actual MFR can be calculated from the reduced MFR m_{turb} by

$$m_{turb} = f_{turb}\left(\frac{P_{in}}{P_{out}}, \frac{N_{turbo}}{\sqrt{T_{in}}}\right)\frac{P_{in}}{\sqrt{T_{in}}} \tag{2.4}$$

and

$$\eta_{turb} = f'_{turb}\left(\frac{P_{in}}{P_{out}}, \frac{N_{turbo}}{\sqrt{T_{in}}}\right) \tag{2.5}$$

2. **Compressor mapping:** Compressor maps, f_{comp} and f'_{comp}, in Equations (2.6) and (2.7) are used to calculate the compressor pressure ratio and thermal efficiency η_{comp} based on reduced MFR m_{comp} and reduced turbo shaft speed by

$$\frac{P_{out}}{P_{in}} = f_{comp}\left(\frac{m_{comp}\sqrt{T_{in}}}{P_{in}}, \frac{N_{turbo}}{\sqrt{T_{in}}}\right) \tag{2.6}$$

and

$$\eta_{comp} = f'_{comp}\left(\frac{m_{comp}\sqrt{T_{in}}}{P_{in}}, \frac{N_{turbo}}{\sqrt{T_{in}}}\right) \tag{2.7}$$

3. **Temperature calculation:** The outlet temperature of the turbine or compressor can be calculated based upon:

$$\frac{T_{out}}{T_{in}} = \left(\frac{P_{out}}{P_{in}}\right)^{\frac{(\kappa-1)}{\kappa}} \tag{2.8}$$

Notice that Equation (2.8) assumes isentropic gas expansion and the compressing process for either turbine or compressor. However, the actual physical process is not isentropic, leading to more enthalpy remaining in the gas due to thermal efficiency, which makes the actual outlet temperature higher than that given by Equation (2.8), but the difference is relatively small. Simulation results presented in [15] show an acceptable correlation between GT-Power simulation results and the temperature calculated using Equation (2.8). Therefore, this assumption is acceptable.

4. **Turbine power calculation:** The power generated by the turbine, denoted as E_{turb}, is calculated by

$$E_{turb} = m_{turb}C_p\eta_{turb}T_{in}\left[1-\left(\frac{P_{out}}{P_{in}}\right)^{\frac{\kappa-1}{\kappa}}\right] \tag{2.9}$$

5. **Compressor power calculation:** The power required to drive the compressor, denoted as E_{comp}, is calculated by

$$E_{comp} = m_{comp} C_p \frac{1}{\eta_{comp}} T_{in} \left[\left(\frac{P_{out}}{P_{in}} \right)^{\frac{\kappa-1}{\kappa}} - 1 \right] \qquad (2.10)$$

6. **Power balance on turbocharger shaft:** The power balance on the turbocharger shaft is calculated by

$$E_{turb} - E_{comp} = J_{turbo} N_{turbo} \frac{dN_{turbo}}{dt} \qquad (2.11)$$

where J_{turbo} is the rotational inertia of the turbocharger shaft.

2.2.2 Crank-Based One-Zone SI Combustion Model

This subsection presents the mathematic model of the spark ignition (SI) combustion model based on the one-zone assumption of the in-cylinder gas-fuel mixture. Multizone (two-zone and three-zone) combustion models can be found in [16, 17].

2.2.2.1 Crank-Based Methodology

The purpose of the combustion process modeling is to correlate the trapped in-cylinder gas properties, such as air-to-fuel ratio, trapped EGR gas mass, and in-cylinder gas pressure and temperature, to the combustion characteristics, such as misfire, knock, burn duration, and indicated mean effective pressure (IMEP). The developed combustion model needs to be combined with the mean value air handling system model to form the entire engine model used for model-based control strategy development and validation. Note that the combustion model also needs to meet the real-time HIL simulation requirements.

Some combustion-related parameters in the combustion model need to be updated every crank degree, such as in-cylinder gas pressure and temperature, while the others, such as IMEP and air-to-fuel ratio, are updated once every engine cycle at the given crank position for individual cylinders. The latter reflects cycle-to-cycle dynamics of the combustion process. Overall, they are all discrete functions of engine crank position θ_i, which is different from the mean value model presented in the previous subsection, where the parameters are continuous functions of time t.

There are many motivations for using the crank-based modeling approach. The first is due to the fact that most combustion characteristics are usually functions of the crank angle, such as burn duration and peak pressure location, and the second is that the entire combustion process is divided into several combustion phases associated with certain events as a function of crank position. As shown in Figure 2.1, these events are intake valve closing (IVC), spark ignition timing (ST), exhaust valve opening (EVO), exhaust valve closing (EVC), and intake valve opening (IVO). The in-cylinder behaviors (such as pressure, temperature, etc.) are modeled differently during each combustion phase that is defined between two combustion events.

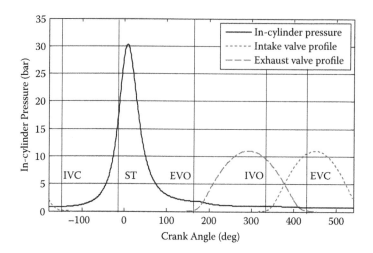

FIGURE 2.1
SI combustion-related events and phases.

The crank-based modeling approach has its own limitations, too. During the real-time simulation the entire model needs to be executed within the time period associated with the desired update period (for example, one crank degree). This leads to a very short computational time window at high engine speed. For instance, at 3000 rpm of engine speed one crank degree corresponds to 56 µs, while at 6000 rpm it reduces to only 28 µs. If the upper limit of engine speed is set to 6000 rpm, in order to avoid the overrun during the real-time simulation, the computation of the combustion model must be completed within 28 µs. This limits the complexity of the combustion model.

In Figure 2.1, the combustion phase starts from ST and ends with EVO. The gas exchange process from EVO to IVC and the compression process from IVC to ST are also important to the combustion process, since the gas-fuel mixture is prepared during these two phases.

2.2.2.2 Gas Exchange Process Modeling

1. **EVO to IVO:** After the exhaust valve is opened, the in-cylinder gas entropically expands in the engine cylinder, exhaust ports, and manifold, and this is called the gas exhaust phase. Simply assume the in-cylinder pressure in this phase equals the exhaust manifold absolute pressure (EMP) P_{EM}

$$P(\theta_i) = P_{EM} \tag{2.12}$$

where θ_i is the current crank position. The in-cylinder gas temperature is calculated by

$$T(\theta_i) = T(\theta_{EVO}) \cdot \left[\frac{P_{EM}}{P(\theta_{EVO})} \right]^{\frac{\kappa-1}{\kappa}} \tag{2.13}$$

where $T(\theta_{EVO})$ and $P(\theta_{EVO})$ are the temperature and pressure at the crank position of exhaust valve closing, and they can be derived from the combustion phase.

2. **IVO to EVC:** This phase is usually called valve overlap phase. During this phase the intake valve starts to open while the exhaust valve is closing. The opening of both valves makes the flow dynamics more complicated and difficult to model. For simplicity, assume the in-cylinder gas pressure equals the mean value of the pressures in exhaust manifold and intake manifold. That is,

$$P(\theta_i) = \frac{P_{EM} + P_{IM}}{2} \tag{2.14}$$

where P_{IM} is the intake manifold absolute pressure. The gas temperature can be calculated the same as in the last phase:

$$T(\theta_i) = T(\theta_{IVO}) \left[\frac{P_{EM} + P_{IM}}{2P(\theta_{IVO})} \right]^{\frac{\kappa-1}{\kappa}} \tag{2.15}$$

In addition, at EVC the residual gas mass is calculated based on ideal gas law as follows:

$$M_r = \frac{P(\theta_{EVC})V(\theta_{EVC})}{T(\theta_{EVC})R} \tag{2.16}$$

3. **EVC to IVC:** During this phase fresh air is trapped inside the engine cylinder. The in-cylinder pressure is mainly influenced by intake manifold pressure, but not always equal to it. It can be calculated by

$$P(\theta_i) = P_{IM}\eta_{IN} \tag{2.17}$$

where η_{IN} is actually the volumetric efficiency of the intake process, and it is a function of engine speed N_e and engine load P_{IM}. In-cylinder gas temperature is calculated by

$$T(\theta_i) = T(\theta_{EVC}) \left[\frac{P_{IM}\eta_{IN}}{P(\theta_{EVC})} \right]^{\frac{\kappa-1}{\kappa}} \tag{2.18}$$

Additionally, the total mass of in-cylinder gas mixture for the compression and combustion phases is calculated at IVC, also based on the ideal gas law, as follows:

$$M_t = \frac{P(\theta_{IVC})V(\theta_{IVC})}{T(\theta_{IVC})R} = \eta_{IN} \frac{P_{IM}V(\theta_{IVC})}{T(\theta_{IVC})R} \tag{2.19}$$

4. **IVC to ST:** This phase is the compression phase without combustion. The governing equations of this phase are also based on the isentropic law of ideal gas. They are

$$P(\theta_i) = P(\theta_{i-1}) \left[\frac{V(\theta_{i-1})}{V(\theta_i)} \right]^{\kappa} \tag{2.20}$$

and

$$T(\theta_i) = T(\theta_{i-1}) \left[\frac{V(\theta_{i-1})}{V(\theta_i)} \right]^{(\kappa-1)} \tag{2.21}$$

where $V(\theta_i)$ is the cylinder volume at crank position $V(\theta_i)$ and is calculated by

$$V(\theta_i) = \left[\frac{1}{2} + \frac{1}{r-1} + \frac{L}{S} - \frac{\cos(\theta_i)}{2} - \sqrt{\frac{L^2}{S^2} - \sin^2(\theta_i)} \right] \frac{\pi B^2 S}{4} \tag{2.22}$$

Note that r is compression ratio, L is connecting rod length, S is piston stroke, and B is piston bore. The values of the sample parameters can be found in [15].

2.2.2.3 One-Zone SI Combustion Model

In the SI combustion model, the start of combustion is initiated by the spark ignition, which can be controlled at any desired crank position defined as spark timing (ST). After ST the mass fraction burned of trapped fuel can be represented by an S-shaped Wiebe function [19] as

$$x(\theta_i) = 1 - \exp\left[-a \left(\frac{\theta_i - \theta_{ST}}{\Delta\theta_{SI}} \right)^{m+1} \right] \tag{2.23}$$

where $\Delta\theta_{SI}$ is the predicted burn duration of the SI combustion mode (a calibration parameter of engine speed, engine load, often represented by the manifold air pressure (MAP), and coolant temperature), and m is the Wiebe exponent ($m = 2$ was used in the model). Coefficient a depends on how the burn duration $\Delta\theta_{SI}$ is defined. In case $\Delta\theta_{SI}$ is defined as the duration of the 10% to 90% MFB, a can be calculated by

$$a = \left\{ \left[-\ln(1-0.9) \right]^{\frac{1}{m+1}} - \left[-\ln(1-0.1) \right]^{\frac{1}{m+1}} \right\}^{m+1} \tag{2.24}$$

Assuming both burned and unburned gases are evenly mixed in one zone, the SI combustion process is simplified into a heat transfer with a volume change process of the entire in-cylinder gas. The in-cylinder gas temperature can be calculated by

$$T(\theta_i) = T(\theta_{i-1}) \left[\frac{V(\theta_{i-1})}{V(\theta_i)} \right]^{(\kappa-1)} + \frac{\eta_{SI} M_f H_{LHV} \left[x(\theta_i) - x(\theta_{i-1}) \right] - Q(\theta_i)}{M_t C_v} \tag{2.25}$$

where η_{SI} is the function of engine speed and load, calibrated by matching the calculated IMEP with that given by GT-Power, and Q represents the heat transfer between the in-cylinder gas and cylinder inner surface. Only convection was considered in the model, since for a gasoline engine the heat transfer due to radiation is relatively small in comparison with the convective heat transfer [20]. The Woschni correlation model [21, 22] is used to calculate the heat transfer term:

$$Q(\theta_i) = A_c h_c [T(\theta_{i-1}) - T_w] \tag{2.26}$$

where h_c is called the Woschni correlation, and it can be written as

$$h_c = \frac{c B^{m-1} P^m w^m T^{0.75-1.62m}}{N_e} \tag{2.27}$$

The coefficients c and exponent m in Equation (2.27) can be used to correlate the simulation results to the experimental data or GT-Power simulation results; $c = 0.54$ and $m = 0.8$ were found to have good correlation for the model in the result presented in Chapter 5.

There are two terms on the right-hand side of Equation (2.25). The first term represents an isentropic compressing or expanding process, while the second term calculates the temperature rise due to the heat transfer during the combustion. Therefore, the complicated thermodynamic process of the combustion is simplified into an isentropic volume change process without heat exchange in one crank degree period and the heat absorption from combusted fuel without volume change in an infinitely small time period. Based on the updated gas temperature from Equation (2.25), the gas pressure can be calculated by applying ideal gas law to the in-cylinder gas as follows:

$$P(\theta_i) = P(\theta_{i-1}) \cdot \frac{V(\theta_{i-1})}{V(\theta_i)} \cdot \frac{T(\theta_i)}{T(\theta_{i-1})} \tag{2.28}$$

2.2.3 Combustion Event-Based Dynamic Model

Due to the cycle-by-cycle combustion event, certain engine dynamics need to be modeled event by event such as fuel injection process and exhaust gas recirculation (EGR).

2.2.3.1 Fueling Dynamics and Air-to-Fuel Ratio Calculation

The engine system could be equipped with port fuel injection (PFI), direct injection (DI), or both PFI and DI systems. Since the fuel injected by the DI fuel system is trapped in the cylinder directly and will not affect the fueling quantity for the next cycle, the DI fuel injection dynamics is normally ignored. For the PFI fuel injection system, the wall-wetting phenomenon of the PFI fuel spray on the intake port and the back of the intake valve introduces cycle-to-cycle dynamics and affects the engine transient performance significantly, and it needs to be modeled in the engine model.

The wall-wetting phenomenon of the PFI fuel injection can be described in such a way that only part of the injected fuel ($\beta \cdot M_{inj}$, $0 < \beta < 1$) enters the cylinder while the rest of the fuel (($1 - \beta$) $\cdot M_{inj}$) remains on the surface of the intake port and the back of intake valves. Then the total fuel mass flow into the cylinder consists of the fuel directly injected into the cylinder and the fuel vapor ($\alpha \cdot M_{res}$, $0 < \alpha < 1$) from fuel mass stored on the intake port and the back of the intake valves from previous injection. The wall-wetting phenomenon leads to the most important dynamics in PFI fuel mass calculation, which affects engine transient performance significantly [18]. The governing equation of the wall-wetting dynamics can be expressed as

$$
\begin{cases}
M_{fuel}[k] = \alpha \cdot M_{res}[k-1] + \beta \cdot M_{inj}[k] \\
M_{res}[k] = (1-\alpha) \cdot M_{res}[k-1] + (1-\beta) \cdot M_{inj}[k]
\end{cases}
\tag{2.29}
$$

where k is an index representing the engine cycle number (k indicates current engine cycle and $k-1$ the last engine cycle), M_{fuel} is the quantity of fuel mass flowed into the cylinder, M_{res} is the quantity of fuel mass left on the port and the back of intake valves, M_{inj} is the amount of fuel injected by the PFI injector at the given engine cycle, and coefficients α and β are functions of engine coolant temperature, engine speed, and load.

The engine gas exchange behavior introduces dynamics to the air-to-fuel ratio calculation too, since a substantial portion of the burned gas remains inside the cylinder, especially at low load. This gas fraction carries the air-to-fuel ratio of the previous engine cycle to the current one. Therefore, the air-to-fuel ratio can be modeled cycle-by-cycle below:

$$
\lambda[k] = \frac{\lambda_f[k]\big(M_t[k] - M_r[k-1]\big) + \lambda[k-1] \cdot M_r[k-1]}{M_t[k]}
\tag{2.30}
$$

where λ is the normalized air-to-fuel ratio of the gas mixture inside the engine cylinder after IVC, λ_f is the normalized air-to-fuel ratio of the fresh charge in the current cycle and it is defined as

$$
\lambda_f[k] = \frac{M_t[k] - M_r[k-1]}{M_{fuel}[k](1+\sigma)}
\tag{2.31}
$$

and σ is stoichiometric air-to-fuel ratio of the fuel.

2.2.3.2 Engine Torque and Crankshaft Dynamic Model

The equations presented in the last subsections provide a complete cycle profile of in-cylinder gas pressure. Based on this pressure profile and the cylinder volume profile, the engine IMEP can be calculated by a simple digital integration:

$$
P_{IMEP} = \frac{1}{V_d} \cdot \sum_{i=0}^{i=719} \big\{ P(\theta_i) \cdot \big[V(\theta_i) - V(\theta_{i-1}) \big] \big\}
\tag{2.32}
$$

where V_d is the cylinder displacement and

$$
V_d = V(\theta_{BDC}) - V(\theta_{TDC})
\tag{2.33}
$$

At last engine torque output is calculated by

$$
T_e = \frac{60n(P_{IMEP} - P_{FMEP})V_d}{2\pi N_e}
\tag{2.34}
$$

where P_{FMEP} is the friction mean effective pressure and n is the quantity of engine cylinders.

Based upon Newton theory, assuming a rigid crankshaft, it can be derived as

$$\frac{dN_e}{dt} = \frac{60}{2\pi} \frac{T_e - T_l}{J_e} \qquad (2.35)$$

where J_e is the rotational inertia of the engine crankshaft; T_e and T_l are the engine brake and load torques, respectively. Note that simulations T_l can be generated by an engine dynamometer model controlled by a proportional-integral-derivative (PID) feedback controller to maintain the desired engine speed.

2.3 Valve Actuation System

2.3.1 Valve Actuator Design

Poppet-type intake and exhaust valves are widely used to control the fresh charge and exhaust gas exchange dynamics during the intake and exhaust strokes of the internal combustion engine (ICE). The valves are actuated with one or two camshafts that are connected to the crankshaft mechanically. There are mainly three different arrangements for the valve actuation system. The direct acting system has the cam lobe in contact with the follower and the engine valve in a vertical arrangement. The roller finger follower system uses a lever to actuate the valve, and the cam lobe is in contact with a roller that is mounted on the lever between the valve and the pivoting point. The pushrod system installs the camshaft in the valley of the engine and uses a pushrod to actuate the valve through a lever. The roller finger follower system has the smallest effective mass, while the pushrod system is heavier than the other two systems due to the long connecting rod. The roller finger follower system also has less friction due to the rolling contact. The direct acting system has the largest friction for its sliding friction. The pushrod system is more suitable for low- to medium-speed operation, and the direct acting system is capable of high-speed engine operation. From the packaging perspective, the pushrod system is able to reduce the overall height of the engine since the camshaft is housed in the valley of the engine.

A conventional valvetrain with fixed valve motion prevents real-time optimization of the air management system. Flexible intake or exhaust valve motions can greatly improve the fuel economy, emissions, and torque output performance of the internal combustion engine. Flexible valve actuation can be achieved with mechanical (cam-based), electromagnetic (electromechanical), electrohydraulic, and electropneumatic valvetrain mechanisms. The cam-based mechanisms offer limited flexibility of the valve event and are designed as multiple-step devices or continuously variable devices. The multistep cam mechanism [23], for example, allows switching between two (or three) discrete cams. The cam phasing mechanism [24, 25] allows the intake or exhaust cams to be continuously phase shifted, however, without the flexibility of changing the valve lift or duration. The variable valve lift system [26] has incorporated a combination of variable cam phasing with a continuously variable valve lift mechanism, which provides significant flexibility, but at relatively high cost and complexity. A fully flexible valve actuation system, often referred to as camless valvetrain, includes electromagnetic (electromechanical), electrohydraulic, and electropneumatic systems. The electromagnetic systems [27] are able to generate flexible valve timing and duration. These devices, however, generally have high valve seating velocity and are limited by the inherent fixed valve lift operation. The electrohydraulic systems [28–32] also provide fully flexible control of the valve lift events. For these systems, digital or proportional valves are used to control the hydraulic fluid to actuate the engine valve. The potential issues with the electrohydraulic systems are energy consumption and reliability of generating a repeatable valve profile over the life cycle of the engine. Electropneumatic systems [33, 34] employ pneumatic actuators to drive the engine valve. Potential issues with the electropneumatic systems include low power density and compressibility of air. Due to the above challenges, currently there is no mass-produced fully flexible valve actuation system on the market.

Motivations for developing fully flexible valve actuation (FFVA) systems come from three areas. First, the FFVA system offers significant fuel economy benefits, lower emissions, and better torque output performance. Second, the FFVA systems could enable various engine operating strategies such as nonthrottling load control, cylinder deactivation, internal exhaust gas recirculation (EGR) control, homogeneous charge compression ignition, switching or combining engine operation modes, etc. Third, the FFVA system could also provide a common platform that delivers the functions of throttle, EGR, cam phaser, cylinder deactivation, port deactivation, two-step cam, and continuously variable lift systems. This would reduce the development time and cost compared with separately developed systems.

Besides the above production-oriented systems, laboratory FFVA systems have also been developed [35–39]. These laboratory FFVA systems have been used to explore various advanced combustion concepts, and the results have shown significant fuel economy, emissions, and performance improvement. More importantly, from the valve actuation perspective, they demonstrate the feasibility of operating FFVA systems safely and reliably in the laboratory environment.

2.3.1.1 Challenges for Developing FFVA Systems

A production-viable FFVA system must be able to generate precise and robust valve motion with high efficiency and simplified control. In this section we will outline the technical challenges for developing FFVA systems and their implications to system design and control. While the challenges are common to any type of FFVA systems, including electromagnetic, electrohydraulic, and electropneumatic systems, we will focus on the electrohydraulic system in this section.

2.3.1.1.1 Precise and Robust Motion

For fully flexible valve actuation systems, without proper planning and control, it is possible to have mechanical interference between the piston and valves or between the engine valves. So precise valve motion is required to prevent mechanical interference. As the valve lift profile controls the airflow and residual level inside the combustion chamber, it becomes necessary to have accurate and consistent valve lift profiles to achieve high-performance engine operation. This consistency requirement includes both cycle-to-cycle and cylinder-to-cylinder repeatability of the valve lift profiles. The impact speed during the valve closing event is called seating velocity. Acceptable valve seating velocity is required to minimize valvetrain noise and maximize valve and valve seat durability.

2.3.1.1.2 Energy Efficiency

The FFVA system must be energy-efficient so that it can not only improve combustion efficiency, but also reduce the amount of energy required to actuate the valvetrain system. For electrohydraulic systems, pump efficiency, throttling of the high-pressure fluid, and parasitic losses due to leakage, etc., are the major sources of energy loss. For conventional servo and proportional valves, precise flow control is realized by throttling fluid with a variable orifice at the cost of energy efficiency. So throttle-less operation is required to reduce energy consumption.

2.3.1.1.3 Simplified Control

With all the flexibilities enabled by the FFVA system, it also comes with the increased complexity of the control system. The timing (phase), lift, and duration of the valve event can be adjusted on the fly depending on the load and speed conditions of the ICE. For a production-viable system, it must require minimum calibration effort to ensure short development time and system robustness.

2.3.1.2 System Design

To address the above technical challenges, we need a precise and robust flow control mechanism that is capable of throttle-less operation with simplified control. Inspired by this requirement, we present an electrohydraulic valve actuation concept with the internal feedback mechanism [40–42] that relays the engine valve motion to the control spool position via a built-in feedback channel to achieve precise and efficient valve motion with little external control effort.

The system is shown in Figure 2.2 [41]. At the beginning of a valve operation cycle, the supply valve {1} is in the deenergized state. Hence, the entire system is connected to the low-pressure reservoir. The only force acting on the hydraulic piston {2} is the spring force,

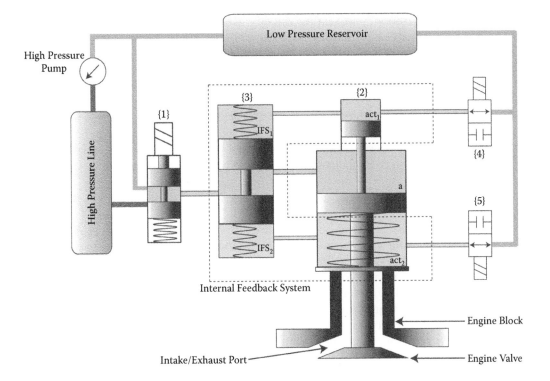

FIGURE 2.2
Diagram of the valve actuation system.

which holds the engine valve in the closed position. Both on-off valves {4 and 5} are open, and hence all the feedback chambers are at the low pressure, and the spool of the feedback regulator {3} is balanced in the center position. To open the engine valve, the system is connected to the high-pressure line by energizing the supply valve. When the pressure in the actuation chamber overcomes the spring force, it accelerates the hydraulic piston downward. When the engine valve reaches a desired lift, the bottom on-off valve {5} is shut off, which restricts the fluid flow from the bottom feedback chamber to the reservoir and diverts it into the bottom chamber of the feedback regulator, and thus deflects the spool upward. As the spool moves upward, the orifice area of the spool valve decreases and thus restricts the fluid flow through it. This reduces the pressure in the actuation chamber and causes the hydraulic piston to decelerate. The synchronized motion between the hydraulic piston and the spool valve will continue until flow to the actuation chamber is completely shut off by the spool and the piston comes to a stop smoothly. Hence, by controlling the timing of the bottom on-off valve, we can control the maximum lift of the engine valve. To close the engine valve, the supply valve is first deenergized. The bottom on-off valve is then opened, which causes the spool of the regulator to return to the center position, and thus connects the actuation chamber to the low-pressure reservoir. The spring force becomes dominant, and hence moves the hydraulic piston upward, which starts to close the engine valve. The top on-off valve {4} is closed when the engine valve is near the seat. By the same principle, as explained previously, the spool is deflected in the downward direction, which decreases the orifice area and thus restricts the flow out of the actuation chamber. The actuation chamber pressure thus increases, which gradually decelerates the engine valve until it lands on the valve seat with a desired seating velocity. Hence, by just controlling the timing of the three valves (the solenoid valve and the on-off valves), the valve timing, lift, duration, and seating velocity can all be controlled precisely. In addition to the effective valve motion control, this system has another advantage. The internal feedback mechanism is a very stiff hydromechanical system and has a very fast response when compared with electromechanical feedback loops. It can be engaged at the last moment possible and would thus lead to throttle-free operation during a major portion of the engine valve cycle. This leads to a minimization of throttling losses, and hence results in a decrease in power consumption.

Detailed dynamics models of the system have been built. The comparison between the model prediction and experimental results is shown in Figure 2.3 [41]. The results clearly demonstrate the precise control of valve lift, seating velocity, and duration.

2.3.2 Valve Actuator Model and Control

For fully flexible valve actuation systems, since there is no mechanical link between the crankshaft and the engine valve, feedback control is critical to achieve precise valve motion. Control system design for FFVA systems has been explored by a number of researchers. Richman and Reynolds [35] presented the development of an electro-hydraulic valve actuation system. An analog controller was used to regulate the valve motion. Performance degradation was observed at high engine speed due to limited bandwidth of the controller. Anderson et al. [43] presented an adaptive peak lift control for an electrohydraulic system. Valve event consistency was affected by the nonlinearity of the device and slow response of the solenoid valve. Misovec et al. [44] presented the digital valve technology applied to the control of a hydraulic valve actuator. Optimal control and proportional control were applied to drive the hydraulic valve to track a 1 Hz sinusoidal curve. Hoffmann and Stefanopoulou [45] reported simulation results

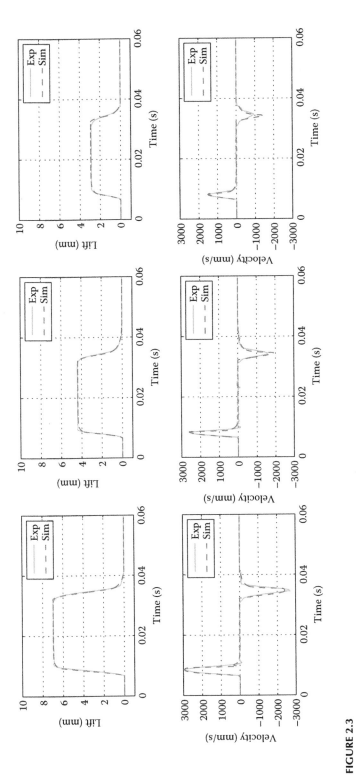

FIGURE 2.3
Comparison between model simulation and experimental results.

for the iterative learning control of an electromagnetic valve actuator to handle the gas flow force to improve system performance. Tai and Tsao [46] presented the modeling and control of an electromagnetic valve actuator. An LQ optimal control was used to achieve soft seating capability.

In this section we present the model and control for a FFVA system [47] to precisely track a desired valve profile that is generated electronically in real time based on the engine operating condition. Due to the four-stroke motion of the ICE, the valve profile is periodic to the rotational angle of the engine. With constant engine speed, the valve profile is also periodic in the time domain, and the lift, phase, and duration transients can be realized using robust repetitive control [48, 49]. When the engine speed varies, the period of the valve profile changes in real time. This phenomenon poses a fundamental challenge to the transient control problem and repetitive control cannot be applied anymore. To overcome this challenge, we use a new valve profile consisting of a periodic portion and a dwell portion with time-varying duration. Robust repetitive control is then applied to the periodic portion, and proportional plus integral and derivative control is applied to the dwell portion. These two controls are switched in real time to achieve asymptotic valve profile tracking performance. To demonstrate the effectiveness of the proposed control method, we show real-time valve lift profiles used to explore homogeneous charge compression ignition (HCCI) combustion at different engine operating conditions.

2.3.2.1 System Hardware and Dynamic Model

As shown in Figure 2.4 [47], the FFVA system includes a hydraulic pump, accumulators, servo valves, hydraulic pistons, linear variable differential transducers (LVDTs), encoders, signal conditioners, power amplifiers, and the control system. The hydraulic pump

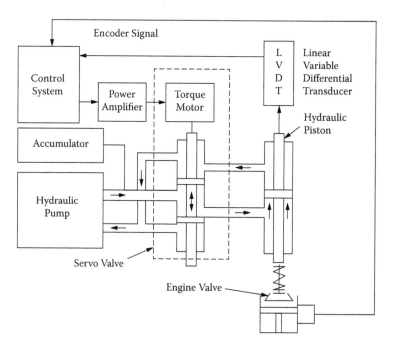

FIGURE 2.4
Block diagram of the FFVA system.

generates high-pressure fluid, and the gas bladder accumulator filled with N_2 is used to minimize supply pressure fluctuations. The servo valve dispenses the fluid in real time according to the control input. The hydraulic fluid that discharges from the servo valve drives the hydraulic piston, which actuates the engine valve. The hydraulic piston is supported by a hydrostatic bearing on each end and has an active cross section area of 77.42 mm² and a maximum travel of 34.34 mm. The LVDT is mounted on top of the piston rod to detect its displacement, and output of the LVDT is conditioned and sent to the control system.

The primary objective of the control system is to ensure precise and flexible valve motion for both steady-state and transient operations. Depending on the specific engine architecture, the clearance between the engine valve and the piston could be extremely small, such as a few crank angle degrees or 1 mm. So precise valve motion is critical for the FFVA system not only because accurate valve timing is required, but more importantly, it prevents mechanical interference between the valve and the piston. The overall control system block diagram is shown in Figure 2.5 [47]. The outer loop is engine control, which generates the desired valve lift profile based on the engine operating condition. The inner loop is valve profile tracking control, which ensures the actual valve motion tracks the desired profile. In this section we focus on the valve profile tracking control.

To precisely track the desired valve profile, we first characterize its dynamics as a function of engine speed and then examine the implications to system dynamics and control design. Due to the four-stroke motion of the ICE, the valve motion is periodic with respect to the rotational angle of the engine. At constant engine speed, the valve profile is also periodic in time domain. With varying engine speed, the period of the valve profile changes in real time and becomes cyclic and aperiodic.

First let's look at the valve motion at constant engine speed. Since the valve profile is periodic in the time domain, we can represent it using the following Fourier series:

$$x(t) = \frac{a_0}{2} + a_1 \cos \omega t + b_1 \sin \omega t + a_2 \cos 2\omega t + b_2 \sin 2\omega t + \cdots \tag{2.36}$$

where ω corresponds to the engine valve cycling frequency, and a_i and b_i are the coefficients of the Fourier series. For periodic signals, the Fourier transform coefficients are only non-zero at discrete frequency points: the first- and higher-order harmonics as shown in (2.36).

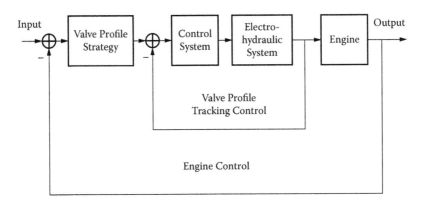

FIGURE 2.5
Control block diagram of the overall system.

So to track the valve profile $x(t)$ precisely in the time domain is equivalent to precise tracking in the frequency domain at those discrete frequency points. Then the question becomes how many harmonics need to be tracked to ensure precise tracking performance. Assume the valve actuation system is an ideal all-pass system with a cutoff frequency, which means the system will pass any signal up to its bandwidth without any amplitude and phase distortion. If the input to the system is $x(t)$ (desired valve profile), the output of the system (the actual valve position) will be

$$y(t) = \frac{a_0}{2} + a_1 \cos \omega t + b_1 \sin \omega t + \cdots + a_k \cos k\omega t + b_k \sin k\omega t \qquad (2.37)$$

where $k\omega$ is the highest harmonic below the cutoff frequency.

Figure 2.6 [47] shows the desired valve profile and the tracking errors for $k = 1$, $k = 5$, and $k = 10$, respectively. Obviously, higher bandwidth produces smaller tracking errors. Please note for computational simplicity, the reference profile used in Figure 2.6 is obtained by using a zero-order hold for a set of measurements of a production cam. The tracking error could be smaller if a first-order hold is used. Table 2.2 summarizes the required bandwidth to pass through 10 harmonics ($k = 10$) at different engine speeds.

If the engine speed varies, the period of the valve motion changes in real time and the valve profile becomes cyclic and aperiodic. For aperiodic signals, the Fourier transform coefficients are nonzero across a continuous frequency spectrum. So to precisely track the aperiodic valve profile in the time domain, we need to track over a continuous frequency spectrum up to the bandwidth specified in Table 2.2 [47]. This requires extremely high

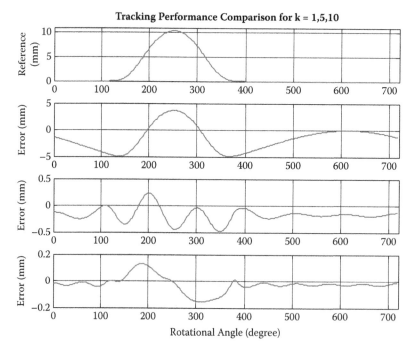

FIGURE 2.6
Desired valve profile and the tracking errors when $k = 1$, 5, 10 (top to bottom).

TABLE 2.2

Required Bandwidth at Different Engine Speeds

Engine Speed (rpm)	Required Bandwidth (Hz)
600	50
1000	83.3
2000	166.7
3000	250
4000	333.3
5000	416.7
6000	500

bandwidth in both the actuator and the control system, which exceeds the capabilities of the state-of-the-art electrohydraulic servo system. To overcome this challenge, as will be shown in the next section, we are able to convert this aperiodic profile tracking problem into a periodic profile tracking problem plus tracking of a constant signal with time-varying duration.

As shown in Table 2.2, regardless of constant or varying engine speed, to precisely track the valve profile in real time, a high-bandwidth response of the FFVA system is required. The ability to achieve a high-bandwidth response depends on a number of factors, including the dynamic response (bandwidth) of the electrohydraulic system and the control system. The bandwidth of the hardware can be limited by the dynamic response of the servo valve, the power amplifier, the feedback position sensor (LVDT), and the hydraulic actuator reciprocating mass. The bandwidth of the control system can also be limited by the sampling rates and the unmodeled dynamics of the plant. As shown in Figure 2.4, the input to the electrohydraulic system is the voltage to the power amplifier, and the output of the system is the LVDT measurement. In this study, we characterize the frequency response of the electrohydraulic system using the swept sine method, where a series of sinusoidal signals from 1 to 1000 Hz is sent to the system. As a result, frequency responses of the four electrohydraulic actuators are shown in Figure 2.7 [47]. The difference between the individual frequency responses is due to the system-built tolerance and calibration variations. Since the control design is model based, a model that captures the system dynamics precisely, including the high frequencies, is essential for achieving the optimal tracking performance and maintaining system robustness. Electrohydraulic systems are essentially nonlinear. However, by incorporating the hydrostatic bearings in the hydraulic piston design, the nonlinear effect of the friction is negligible. Since the maximum stroke for the valve actuator is about 10 mm, we can use a linear model around the operating point and lump the nonlinear effect into the unmodeled dynamics [50]. The discrete-time transfer functions developed for the four intake and exhaust electrohydraulic actuators based on their frequency responses are shown in Equations (2.38) to (2.41). The difference between the experimental data and the model prediction will be treated as unmodeled dynamics.

$$\frac{B_{INTF}(q^{-1})}{A_{INTF}(q^{-1})} = \frac{2.3010 \times 10^{-5} q^{-1} + 8.2900 \times 10^{-4} q^{-2} - 8.4690 \times 10^{-4} q^{-3} + 2.7950 \times 10^{-3} q^{-4} + 9.8410 \times 10^{-4} q^{-5}}{1 - 4.2212 q^{-1} + 8.1133 q^{-2} - 9.8586 q^{-3} + 9.0101 q^{-4} - 6.7874 q^{-5}}$$

$$\frac{-1.2820 \times 10^{-3} q^{-6} - 1.4950 \times 10^{-3} q^{-7} - 3.3410 \times 10^{-4} q^{-8}}{+4.0870 q^{-6} - 1.6750 q^{-7} + 0.3328 q^{-8}}$$

(2.38)

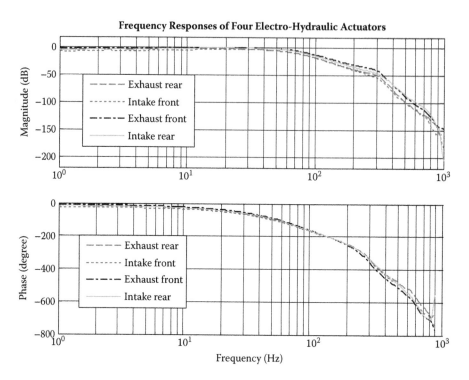

FIGURE 2.7
Frequency responses of the electrohydraulic actuators.

$$\frac{B_{INTR}(q^{-1})}{A_{INTR}(q^{-1})} = \frac{2.1800 \times 10^{-3} q^{-1} - 5.3610 \times 10^{-3} q^{-2} + 7.3120 \times 10^{-3} q^{-3} - 1.0260 \times 10^{-2} q^{-4} + 2.4310 \times 10^{-3} q^{-5}}{1 - 4.2241 q^{-1} + 8.1815 q^{-2} - 9.9051 q^{-3} + 8.7871 q^{-4} - 6.3121 q^{-5}}$$

$$\frac{-8.3210 \times 10^{-4} q^{-6} + 4.7020 \times 10^{-3} q^{-7} - 3.0130 \times 10^{-3} q^{-8}}{+3.6648 q^{-6} - 1.4918 q^{-7} + 0.3027 q^{-8}}$$

$$(2.39)$$

$$\frac{B_{EXHF}(q^{-1})}{A_{EXHF}(q^{-1})} = \frac{9.5230 \times 10^{-5} q^{-1} - 9.2090 \times 10^{-5} q^{-2} - 3.4240 \times 10^{-4} q^{-3} - 1.3550 \times 10^{-3} q^{-4} - 6.5200 \times 10^{-3} q^{-5}}{1 - 3.7302 q^{-1} + 6.2683 q^{-2} - 6.4578 q^{-3} + 4.8969 q^{-4} - 3.2302 q^{-5}}$$

$$\frac{+5.2340 \times 10^{-4} q^{-6} + 1.7060 \times 10^{-3} q^{-7} + 1.0120 \times 10^{-4} q^{-8}}{+1.9011 q^{-6} - 0.8223 q^{-7} + 0.1803 q^{-8}}$$

$$(2.40)$$

$$\frac{B_{EXHR}(q^{-1})}{A_{EXHR}(q^{-1})} = \frac{-1.1290 \times 10^{-4} q^{-1} - 6.6670 \times 10^{-5} q^{-2} + 8.4160 \times 10^{-4} q^{-3} + 2.0860 \times 10^{-3} q^{-4} + 2.8520 \times 10^{-3} q^{-5}}{1 - 3.9259 q^{-1} + 6.6637 q^{-2} - 6.7812 q^{-3} + 5.2700 q^{-4} - 3.8812 q^{-5}}$$

$$\frac{-3.3190 \times 10^{-3} q^{-6} - 1.8920 \times 10^{-3} q^{-7} - 7.9950 \times 10^{-5} q^{-8}}{+2.5466 q^{-6} - 1.1104 q^{-7} + 0.2188 q^{-8}}$$

$$(2.41)$$

where q^{-1} is the one-step delay operator. $A_{INTF}(q^{-1})$, $B_{INTF}(q^{-1})$, $A_{INTR}(q^{-1})$, $B_{INTR}(q^{-1})$, $A_{EXHF}(q^{-1})$, $B_{EXHF}(q^{-1})$, $A_{EXHR}(q^{-1})$, and $B_{EXHR}(q^{-1})$ are the denominators and numerators of the transfer functions for the intake front, intake rear, exhaust front, and exhaust rear actuators, respectively.

2.3.2.2 Robust Repetitive Control Design

At constant engine speed, the valve profile is periodic in the time domain. Robust repetitive control [48, 49] can be applied to achieve the required tracking performance. A key feature of repetitive control is the extremely fast convergence rate of the tracking error due to its high feedback gains at the desired frequency locations. Thus, the lift, duration, and phase transients can all be accommodated using the robust repetitive control. We represent the closed-loop system as follows:

$$y(k) = \frac{B(q^{-1})}{A(q^{-1})} u(k) \tag{2.42}$$

$$u(k) = C(q^{-1})[r(k) - y(k)] \tag{2.43}$$

where k is the discrete-time step index, $u(k)$ and $y(k)$ are the input and output of the electrohydraulic system, respectively, $r(k)$ is desired valve profile, and $A(q^{-1})$ and $B(q^{-1})$ are the actuator models defined in Equations (2.38) to (2.41).

The robust repetitive controller $C(q^{-1})$ has the following form [49]:

$$C(q^{-1}) = M(q^{-1}) \frac{Q(q^{-1})q^{-N}}{1 - Q(q^{-1})q^{-N}} \tag{2.44}$$

where

$$M(q^{-1}) = \frac{A(q^{-1})B^{-}(q)}{B^{+}(q^{-1})b}, \quad b \geq \max_{\omega \in [0,\pi]} |B^{-}(e^{-j\omega})|^2$$

$B^{+}(q^{-1})$ and $B^{-}(q^{-1})$ are the stable and unstable parts of $B(q^{-1})$, respectively,

$$Q(q^{-1}) = \frac{(q + 2 + q^{-1})^n}{4}$$

$n = 1$, and N is the period of the reference signal $r(k)$.

By plugging the controller expressions (2.43) and (2.44) into the plant model (2.42), we have

$$y(k) = \frac{BMQq^{-N}}{A(1 - Qq^{-N}) + BMQq^{-N}} r(k)$$

$$e(k) = r(k) - y(k) = \frac{A(1 - Qq^{-N})}{A(1 - Qq^{-N}) + BMQq^{-N}} r(k)$$

Since the desired valve profile $r(k)$ is periodic with period N, we have $(1-q^{-N})r(k) = 0$, so $\lim_{k\to\infty} e(k) = 0$ if $Q = 1$.

To accommodate the plant unmodeled dynamics and ensure robust stability, we need to compromise between tracking performance and system robustness. Define the unmodeled dynamics as

$$\Delta(e^{-jw}) = \frac{G_0(e^{-jw}) - G(e^{-jw})}{G_0(e^{-jw})} \qquad (2.45)$$

where $G_0(e^{-jw})$ is the nominal plant model defined in (2.42) and $G(e^{-jw})$ is the experimental data obtained from frequency response.

The Q filter defined in (2.44) needs to satisfy the following condition [49] to ensure system robust stability:

$$\frac{1}{|\Delta(e^{-jw})|} \geq |Q(e^{-jw})| \qquad (2.46)$$

Frequency responses of $Q(e^{-jw})$ and $1/\Delta(e^{-jw})$ for the exhaust front actuator are shown in Figure 2.8 [47]; clearly condition (2.46) is satisfied. To illustrate system performance, Figure 2.9 [47] shows the frequency response of the sensitivity function of $e(k)$ with respect to $r(k)$ for the exhaust front actuator with closed-loop control. The notches at those harmonics indicate the corresponding tracking performance. For example, if a 10 Hz sine wave with amplitude 1 is applied to the system, the tracking error would be around 0.0001778. Also, as it is shown, the notch becomes smaller when it moves to the high-frequency range, which illustrates the trade-off between performance and robustness.

When engine speed varies, the period of the valve motion changes in real time. This phenomenon poses a fundamental challenge to the control design for achieving precise tracking. Some advanced control algorithms such as repetitive control cannot be applied

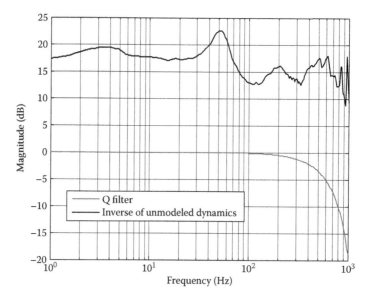

FIGURE 2.8
Frequency responses of the Q filter and the inverse of unmodeled dynamics for the exhaust front actuator.

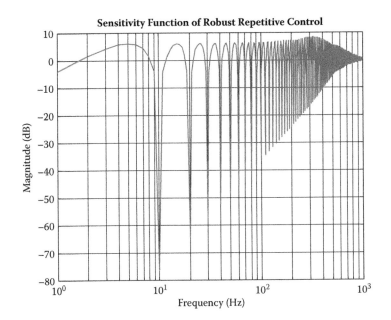

FIGURE 2.9
Frequency response of the sensitivity function for the exhaust front actuator.

anymore under the speed transient. To track those profiles precisely, the FFVA system must have the capability of precise tracking over a continuous frequency spectrum, which usually demands an extremely high bandwidth well exceeding the hardware capability. To overcome this challenge, we convert the aperiodic profile tracking problem into a periodic profile and a constant profile with time-varying duration. Robust repetitive control is then applied to track the periodic profile, and proportional plus integral and derivative control is applied to track the constant profile with time-varying duration. These two controls are switched in synchronization with the valve motion.

To implement the above transient control method, we use a new valve profile as shown in Figure 2.10 [48]. This profile consists of four portions: the seat portion, the opening portion, the lift portion, and the closing portion. As it is shown, the opening and closing profiles of the valve event remain the same in the time domain even when the engine speed changes, but the time duration of the seat and lift dwell portions of the valve event changes with the engine speed. So the opening and closing profiles become periodic in the time domain from one engine cycle to the next independent of the engine speed. The varying period of the valve event is accommodated by the time-varying durations of the lift and seat dwells. We also note that at either end of the opening or closing profile, there is a flat portion to ensure smooth transitions to and from the dwell portion.

Following is the procedure to control the FFVA system to track the above-mentioned profile. The engine valve is at the seat when the operation starts. At the desired opening timing, we transit into the opening portion. Robust repetitive control (2.44) is used to control the FFVA system to track the opening profile. At the end of the opening profile, the engine valve stops at the predetermined lift and enters the lift dwell portion. We then switch from repetitive control to a PID regulator as shown below:

$$u(k) = u_0 + k_P e(k) + k_I \left[\sum e(k) \right] + k_D \left[e(k) - e(k-1) \right] \qquad (2.47)$$

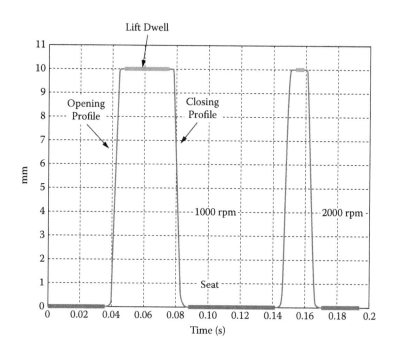

FIGURE 2.10
New valve profile at 1000 and 2000 rpm engine speed.

where u_0 is the initial value and k_P, k_I, and k_D are the proportional, integral, and derivative gains, respectively.

At the desired closing timing, we transit from lift dwell into closing portion. The control system is switched back to the repetitive controller (2.44) from the PID regulator (2.47) to track the closing profile. At the end of the closing portion, the engine valve reaches the seat and enters the seat dwell portion. We switch again the feedback control to the PID regulator.

The above procedure repeats itself for the next valve event regardless of engine speed. The key advantage of the proposed control method is that it provides a unified framework to control the FFVA systems and similar applications for constant or varying engine speed. Another advantage is that the system in fully flexible in terms of lift, phasing, and duration even during the speed transient operation. The underlying principle of the control algorithm is that it converts the control problem from tracking over a continuous frequency spectrum to the tracking at discrete frequency points. Alternatively the recently developed time-varying internal model-based controller [51] can be applied to directly track the time-varying reference signal.

2.3.2.3 Experimental Results

Following are the experimental results for lift, duration, phase, speed, and mode transients.

Lift transient: As shown in Figure 2.11 [47], the desired valve lift was changed from 5.5 to 3 mm after cycle 1, and the FFVA system was able to respond to the change immediately. The maximum tracking error during cycle 2 is about 0.6 mm. The tracking error reduces to less than 0.05 mm at cycle 3. In other words, it takes

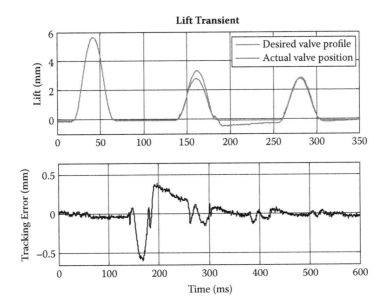

FIGURE 2.11
Transient performance of variable lift control.

only two engine cycles for the transient tracking errors to disappear in the case of a step lift change. Please note that the position signal shown in the figure is the position of the hydraulic piston instead of the engine valve due to the specific location of the sensor (see Figure 2.4). When the position signal is positive, the hydraulic piston and the engine valve are moving together. When the position signal becomes negative, the hydraulic piston separates from the engine valve and continues to retract while the engine valve stops at the valve seat. It is also worth pointing out that the measured position signal itself has a 0.02 mm noise level, so the steady-state tracking errors after the lift transient are mainly due to the position sensor noise.

Duration transient: A step change of 15° in valve duration was tested. As shown in Figure 2.12 [47], the duration of the valve event is increased by 15° in one step in the desired valve profile, and the actual valve position was able to follow the command in the next cycle. The tracking error converges to less than 0.2 mm in two cycles.

Phase transient: A step change of 15° in valve phasing was tested. As shown in Figure 2.13 [47], the phasing of the valve event is advanced by 15° in one step in the desired valve profile, and the actual valve position was able to follow the command in the next cycle. The tracking error converges to less than 0.3 mm in three cycles.

Speed transient: Both bench and combustion tests were conducted to test the FFVA system's speed transient capability. Figure 2.14 [47] shows the bench test results where the FFVA system was running on the engine head without combustion and the engine encoder signal was generated by a simulator. The engine speed was changed from about 1000 to 2000 rpm. The FFVA system was able to adjust the valve event automatically in real time to accommodate the speed transient. Figure 2.15 [47] shows the combustion testing results where HCCI combustion was

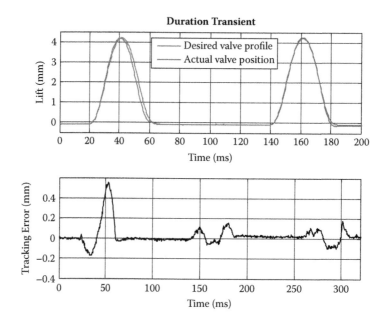

FIGURE 2.12
Transient performance of variable duration control (15° step change).

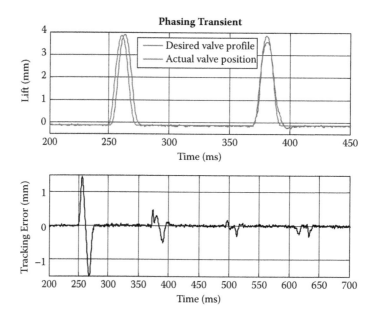

FIGURE 2.13
Transient performance of variable phasing control (15° step change).

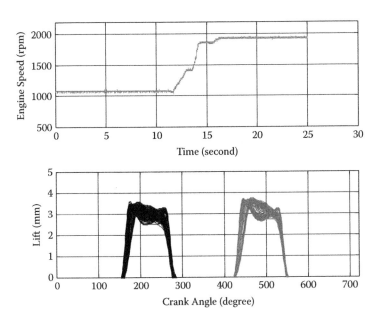

FIGURE 2.14
Speed transient test with simulated encoder signal. Top: Engine speed. Bottom left: Exhaust valve. Bottom right: Intake valve.

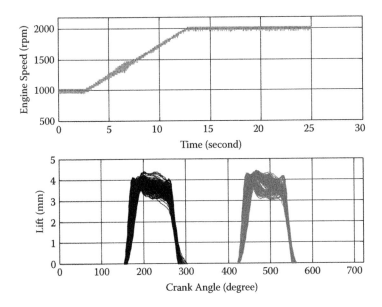

FIGURE 2.15
Speed transient test with HCCI combustion. Top: Engine speed. Bottom left: Exhaust valve. Bottom right: Intake valve.

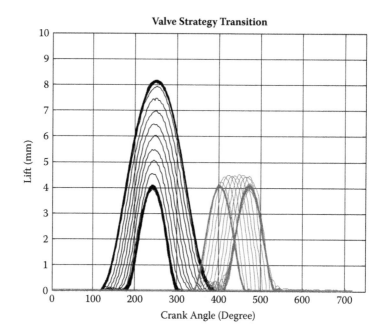

FIGURE 2.16
Valve strategy transition between recompression and nonthrottling load control. Left: Exhaust valve. Right: Intake valve.

conducted during the speed transient and the encoder signal was generated from the encoder sensor on the crankshaft. Again, the FFVA system was able to adjust the valve event in real time based on the engine speed.

Mode transient: Mode transient was tested to accommodate combustion mode switch between HCCI and spark ignition (SI) operations. Figure 2.16 [47] shows the valve strategy change in real time between the recompression valve strategy and the nonthrottling load control valve strategy. The exhaust valve changed from 4 mm to about 8 mm with a wider duration. The intake valve advanced about 50°. We would like to point out that it is important to have precise tracking to perform such a mode transient since the engine valves could get very close to the piston during the transient.

2.4 Fuel Injection Systems

As discussed in Section 2.2, there are two kinds of fuel systems for a gasoline engine: PFI and DI fuel systems. The control aspect of both PFI and DI systems will be discussed in this subsection.

2.4.1 Fuel Injector Design and Optimization

This subsection mainly addresses the DI fuel system design and optimization.

2.4.1.1 PFI Fuel System

The PFI fuel system injects the fuel on the intake port and the back of the intake valves, and during this process, part of the injected fuel flows directly into the cylinder and part of the injected fuel remains on the intake ports and the back of the intake valves. The fuel injector design is mainly to meet the fuel flow requirement to have the desired air-to-fuel ratio and engine output torque.

Figure 2.17 shows the relationship between the fuel injection pulse width and injected fuel, where the dotted line shows the injector knee, and over that region, the injected fuel quantity may not correlate to the injection pulse width consistently. Therefore, the selected fuel injector should not be operated at that region. The fuel injector design or selection is to make sure the fuel injection quantity meets the requirement over the entire engine operational range (speed and load).

2.4.1.2 DI Fuel System

The DI fuel system is quite different from the PFI one. As fuel is injected directly into the engine cylinder, the DI fuel system offers great flexibility to the fuel injection strategy with respect to various engine operation modes. In particular, the fuel-air mixture preparation in the combustion chamber has also been identified as one of the key factors that greatly influence the combustion characteristics of the engine performance [52]. Hence, optimizing the fuel mixture homogeneity and avoiding fuel impingement are key engine design parameters.

When developing combustion systems for DI gasoline engines, it is important to achieve an optimal fuel-air mixture for ignition. Depending upon the combustion chamber configuration and the engine operating modes, the fuel mixture strategy may require different levels of control over key spray characteristics, including spray pattern, cone angle, penetration, and drop size. If the injectors can be designed to offer spray tailoring flexibility, engine designers may utilize the injectors to deliver the specific flow and spray requirements without major compromises and limitations when running the engine at its optimized configuration.

High-speed imaging has evolved as a primary optical diagnostic technique for investigating the characteristics of ultra-fast-motion events. The short time duration between frames and high image quality with good image resolution make high-speed imaging

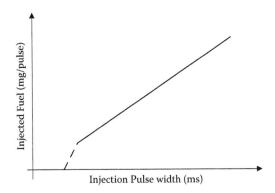

FIGURE 2.17
Injected fuel vs. injection pulse width.

an ideal optical tool to study the highly transient fuel spray characteristics applicable in an engine configuration. Earlier research by Hamady et al. [53] studied the fuel spray characteristics from various injector nozzles using a high-speed imaging system. Using a similar technique, Kawajiri et al. [54] were able to investigate the interaction between spray and air motion in a cylindrical vessel with swirling intake gas motion similar to that in an engine. In addition, high-speed imaging visualization from consecutive cycles was also applied to study fuel distribution, ignition, and combustion characteristics [55–57] under realistic engine speed and load configurations. More recently, Hung et al. [58] combined high-speed imaging with time-resolved laser diffraction to characterize the transient nature of the gasoline pulsing sprays under atmospheric conditions. Transient characteristics such as drop sizing, intrapulse, and pulse-to-pulse interactions throughout and in between consecutive injection cycles were readily resolved.

All fuel spray imaging tests can be performed with a motoring optical engine, where a Mie scattering technique is used to visualize the liquid phase of the fuel dispersion inside the combustion chamber through the quartz cylinder liner wall as well as the quartz piston insert; see [59] for details. The fuel spray can be imaged with a nonintensified high-speed digital video camera. A high repetition rate pulsed copper vapor laser, synchronized with the high-speed camera and the fuel injection timing logic, can be used to illuminate the liquid fuel dispersion. A fiber optics cable can be used to direct the laser pulse through the quartz piston insert into the cylinder; see [59] for details. This arrangement maximized the illumination quality inside the cylinder and minimized most of the secondary scattering from the internal reflection of the quartz wall.

Figure 2.18 shows the comparison of three injector sprays on fuel mixture distribution in the combustion chamber at the engine part load condition of 1500 rpm with a manifold air pressure (MAP) of 45.5 kPa absolute, where injector 40/0 has a 40° spray angle with 0° offset, injector 60/5 has a 60° spray angle with 5° offset, and injector 80/10

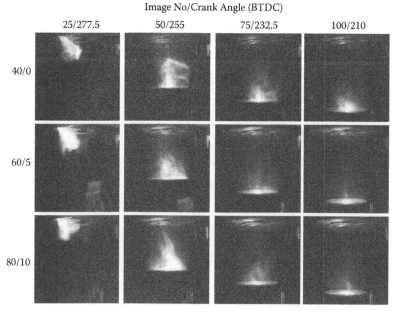

FIGURE 2.18
In-cylinder fuel mixture formation.

has an 80° spray angle with 10° offset. The start of injection (SOI) was set at 300° crank angle (CA) before top dead center (BTDC). With the adjusted cam phasing timing, the intake valve was lifted to about 8 mm at this SOI. The injection pulse width (duration) was set to correspond to lambda equal to 1 (stoichiometric) condition. These images were recorded within an injection cycle at different crank degrees. It is worth mentioning that the injector driver has 1 ms precharge delay, and so it corresponded to the delay of either 9 CAD/ms (at 1500 rpm) or 15 CAD/ms (at 2500 rpm) before the fuel spray was observed at the top of the cylinder. Note that the precharge delay definition can be found in Figure 2.22. The first image of the sequence was shown at 277.5° BTDC, where the initial portion of the spray entering the cylinder was found to be about the same for all three sprays. The intake air did not have much effect on the beginning of the spray. The narrow spray of 40/0 showed a slightly stronger axial penetration along the injector axis into the cylinder. At 255° BTDC, the fuel charge started to show some noticeable differences in the fuel distribution. The 40/0 spray penetrated more directly across the cylinder toward the liner wall, while the 60/5 and 80/10 sprays were moving more toward the central region of the cylinder. They produced very minimal fuel impingement on the opposite side of the liner wall. It also shows that at this SOI timing, the leading portion of the sprays impinged slightly on the piston top. However, as the cycle progressed to 210° BTDC, the fuel distributions among all three sprays were quite similar.

Based on the distinct features depicted in Figure 2.19, it is possible to identify and extract more information on the mixture formation from these images with image processing. Therefore, image processing algorithms were developed in [59] to measure the semiquantitative information, such as the magnitude of fuel spray impingement on cylinder wall and piston top, and fuel-air mixture homogeneity.

For example, to analyze the fuel impingement magnitude on the cylinder wall, a fuel impingement index on the cylinder wall can be defined based upon the illumination intensity of the location (pixel) on the image near the cylinder wall. The methodology of the fuel impingement on the cylinder wall is briefly outlined below. Figure 2.20

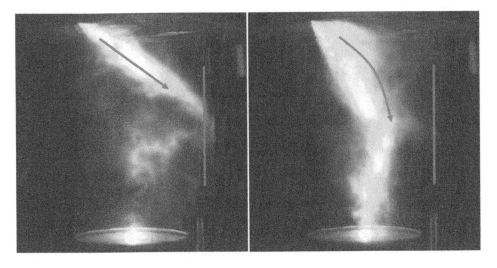

FIGURE 2.19
Location of the fuel impingement analysis (illustrated by the measurement region). Left: 40/0 spray. Right: 80/10 spray.

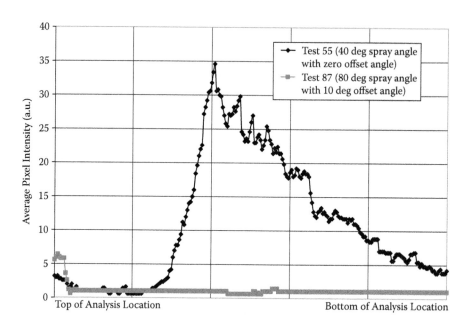

FIGURE 2.20

Comparison of ensemble average intensity on wall impingement when piston is at bottom dead center (BDC).

depicts the measurement areas shown as a gray bar along the cylinder wall where the illumination intensity of each pixel in the image is extracted. The size of this area depends on the image orientation and measurement location of interest. For the analysis of fuel impingement on the cylinder wall, a thin area band was chosen to be 5 pixels (*i*th) wide by 300 pixels (*j*th) long. Then, an average intensity is computed by averaging the pixel intensity across the width (i.e., across the *i*th direction) of the area at each *j*th pixel as follows:

$$\bar{I}_{ave,j} = \frac{\displaystyle\sum_{i=1}^{N} I_{i,j}}{N} \tag{2.48}$$

where $I_{i,j}$ is the intensity of an individual pixel in the measurement area. N is number of pixels along the width, and it is equal to 5 for this analysis.

Figure 2.20 shows the comparison of the average intensity between the two spray patterns along the measurement line. For both sprays, there was no fuel impingement near the top of the cylinder wall. However, it can be seen that for the 40/0 spray, the average intensity along the cylinder wall started to increase abruptly at about one-third of the stroke distance, and it peaked at about halfway on the cylinder. After the peak, the intensity continued to decrease toward the bottom of the cylinder. Fuel impingement was found to spread more on the lower half of the cylinder wall. Conversely, for the wider spray of 80/10, there was almost no impingement of fuel along the cylinder wall. The average intensity remained very minimal and constant along the entire analysis location.

Since the fuel impingement is strongly transient and it is rapidly changing at different crank angles within an engine cycle, an overall fuel impingement index (FII) at a specific crank angle can also be defined based on the ensemble average intensity over the entire location along the jth direction of the measurement domain:

$$FII_{CA} = \frac{\sum_{j=1}^{M} \bar{I}_{ave,j}}{M} \tag{2.49}$$

where M is the number of pixels along the length of the measurement domain and M is equal to 300 for this analysis. This index can be used to track and analyze the extent of fuel impingement at each crank angle degree over the injection cycle. Figure 2.21 shows such a plot of crank angle-resolved fuel impingement index on the cylinder wall. This figure also reveals several useful facts about characteristics such as the sequence and the duration of the fuel impingement. The injection logic pulse started at 270° BTDC when the image sequence was commenced. Taking the injector driver precharge delay into account, the fuel spray entered the cylinder at about 246° BTDC. The spray then propagated directly across the cylinder and started to impinge on the cylinder wall at about 210° BTDC. The impingement index started to increase as the piston continued to sweep downward. For both sprays, even though the peak of impingement was observed to be between 140° and 130° BTDC, the narrow spray with 40/0 resulted in substantially higher cylinder wall impingement than the wider spray of 80/10. Impingement continued to decrease for both sprays as the piston reached about 100° BTDC. Beyond this crank angle, the fuel impingement for both sprays was found to be very minimal.

Similar to any other image analysis methods based on light intensity extracted from the pixels of an image, it is important to realize that the fuel impingement analysis technique mentioned above also requires a consistent light illumination and background in the region

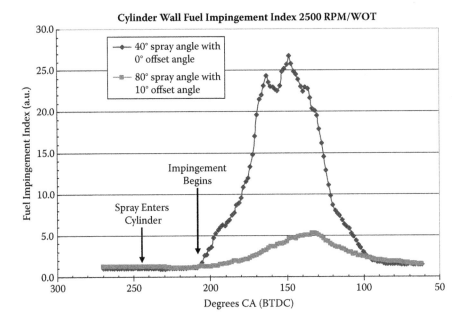

FIGURE 2.21
Crank angle resolved fuel impingement index on cylinder liner wall.

of interest in order to minimize any possible inaccuracy or uncertainty. For example, any fuel droplets populated in the bottom of the liner wall or in the shadow of the piston quartz insert may not be accounted for equally due to the uneven light intensity distribution inside the cylinder. In addition, any residual fuel left behind from previous cycles may remain in the cylinder or be deposited on the walls as the cycle progresses. This could potentially overestimate the quantity and location of the fuel impingement at a particular crank angle. Therefore, a proper background subtraction may be needed to eliminate any contribution of the residual fuel from previous cycles.

Even though this fuel impingement index cannot be used directly to correlate the amount of fuel impinged on the wall, this value indicates the extent as well as the location of the fuel impingement at a specific crank angle within a cycle. Since it is based on the illumination intensity of the pixel, once the images are properly adjusted to correct for any illumination deviation in the imaging setup, it may be useful for comparing other fuel mixing qualities between different conditions.

As a summary, high-speed imaging can be performed to visualize the spray pattern effect on fuel mixture formation as a function of crank angle in a single-cylinder engine for direct injection gasoline applications. With the use of the imaging diagnostics to differentiate the fuel mixing characteristics produced by three different spray patterns, it was found that the spray angle, offset angle, and injector mounting orientation had pronounced effects on the fuel mixture preparation. The fuel mixture inside the combustion chamber was affected more by the spray pattern at full load than on part load condition.

Fuel impingement on cylinder liner walls can also be investigated by using image processing and analysis algorithms. Using high-speed imaging, the transient nature of fuel impingement was resolved as a function of crank angle degree. If a consistent light intensity through the image was ensured, the location and extent of fuel impingement of various spray patterns could be differentiated and compared. Similar image analysis methods may also be applied to evaluate the fuel impingement on the top of the piston. A new injector spray pattern is currently being revised that will not only minimize the fuel impingement on the liner wall and intake valves, but also enhance the overall fuel distribution. These results can be used to correlate the engine combustion and emission performance in the subsequent single-cylinder dynamometer combustion testing. Detailed high-speed visualization of in-cylinder fuel spray and the associated impingement study approach can be found in [59] and [60].

2.4.2 Fuel Injector Model and Control

The fuel injector dynamics for PFI injectors is normally ignored due to relatively fast transient dynamics compared to the wall-wetting dynamics. The main dynamics in the engine model is described in Section 2.2.3.

However, for DI injectors, due to the high fuel pressure, the relationship between injected fuel and commanded injection pulse width is complicated. It was apparent in [61] that there was little uniformity throughout the automotive industry in the use of the term *pulse width*. The lack of uniformity becomes quite important for gasoline DI injection systems, due to the fact that the pulse control strategies can be significantly more complex than those of PFI systems. Many, but not all, DI drivers incorporate a designed driver charge delay (DCD) time between the initiation of the logic pulse from the engine control unit (ECU) and the command pulse of the driver to open the injector. This is normally done to ensure a maximum voltage level of the driver capacitor just prior to actuation. The interrelationships of four key timing traces are illustrated schematically in Figure 2.22.

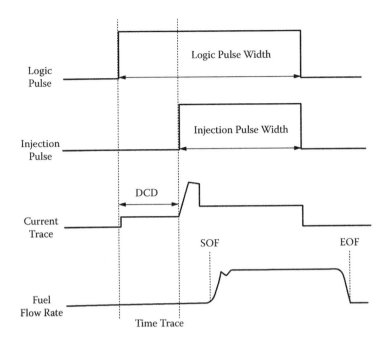

FIGURE 2.22
From logic fuel pulse width to start and end of fuel flow.

In the progression from the ECU logic pulse to the fuel delivery event there are alterations due to a possible DCD, as well as the effects of the individual times of mechanical opening and mechanical closing. The actual fuel delivery time between the first appearance of fuel at the injector tip (start of fuel (SOF)) and the end of fuel (EOF) is not generally equal to either the logic pulse width or the injection pulse width.

2.5 Ignition System Design and Control

Internal combustion (IC) engines are optimized to meet exhaust emission requirements with the best fuel economy. Spark timing is used as one of the optimization parameters for the best fuel economy within given emission constraints. For normal operation, engine spark timing is often optimized to provide minimal advance for the best torque (MBT). On the other hand, engine combustion stability and knock avoidance requirements also constrain engine spark timing within a certain region, called the feasible spark timing region. For certain operational conditions, it is desirable to operate the engine at the borderline of the feasible region continuously. For instance, under certain operational conditions engine MBT timing is located outside of the feasible spark timing region due to the requirement to avoid engine knock. In order to obtain maximum brake torque, it is required to operate the engine at the knock limit (borderline knock limit) of the feasible region. Under different operating conditions, in order to reduce cold-start hydrocarbon (HC) emissions, it is desired to locate the spark timing at the retard limit of the feasible region for fast catalyst light-off. This is due to the desire to maintain a certain level of combustion stability.

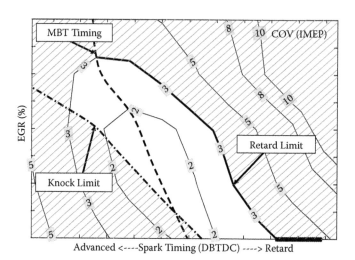

FIGURE 2.23
Feasible region of spark timing.

Figure 2.23 shows a typical spark timing feasible region as a function of exhaust gas recirculation (EGR) for a desired level of combustion stability with coefficient of variation (COV) of IMEP less than 3%, where the thick dashed line represents the engine MBT spark timing, the thick dash-dot line represents the engine advanced (knock) spark limit, and the thick solid line represents the retard spark limit. It can be observed that knock, MBT, and retard limits vary as a function of EGR rate, which makes it difficult to control the optimal spark timing in an open-loop fashion. Further, this feasible region varies in shape with different engine operational and environmental conditions.

In current production applications, MBT timing is an open-loop feedforward control whose values are experimentally determined by conducting spark sweeps at different speed and load points, and at different environmental operating conditions. Almost every calibration point needs a spark sweep to see if the engine can be operated at the MBT timing condition. If not, a certain degree of safety margin is needed to avoid preignition or knock during engine operation. Open-loop spark mapping usually requires a tremendous amount of effort and time to achieve a satisfactory calibration.

Existing knock spark limit control utilizes an accelerometer-based knock sensor for feedback control. Due to the low signal-to-noise ratio, conventional approaches are based on the use of a knock flag signal obtained by comparing the knock intensity signal of a knock sensor to a given threshold. The knock intensity signal is defined as the integrated value, over a given knock window, of the absolute value signal obtained by filtering the raw knock sensor signal using a band-pass filter. This knock flag signal is the input to a dual-rate (slow and fast correction) count-up/count-down engine knock limit controller. The disadvantage of this control scheme is that it continually takes the engine in and out of knock, rather than operating continually at the desired borderline knock limit. In addition, at certain operating points knock observability can be severely compromised by engine mechanical noises such as valve closures and piston slap, which may be picked up by the accelerometer. Such issues result in conservative ignition timing that leads to reduced engine performance.

As discussed before, during a cold start, it is desirable to operate the engine at its retard spark timing limit for minimal HC emissions. The retard spark timing limit is often constrained by engine combustion stability metrics such as COV of IMEP. Due to unavailability of production-ready in-cylinder pressure sensors, the retard spark timing limit is obtained through an offline engine mapping process, leading to conservative calibrations. In addition, to accommodate the range of fuels used throughout a market, this calibration is made even more conservative.

In recent years, various closed-loop spark timing control schemes have been proposed based upon in-cylinder pressure measurements [61–68] or spark ionization current sensing [69–72, 77]. Based upon test data, it has been found that the peak cylinder pressure (PCP) usually occurs around 15° after top dead center (TDC) at MBT timing [69]. The 50% mass fraction burned (MFB) point generally occurs between 8° and 10° after TDC when MBT timing is achieved. The algorithm published in [63] controls PR(10) (normalized pressure ratio of in-cylinder and motoring pressures at 10° after TDC) around 0.55 to obtain the MBT timing.

Due to recent advances in electronics technology, ionization current can be detected at 15 µA with very low background noise. The high quality of an in-cylinder ionization signal makes it possible to derive a linear knock intensity that is proportional to the knock level [74–76]. It can also be processed to derive a metric for combustion quality similar to the COV of IMEP and closeness of combustion to partial burn and misfire limit [76, 82], which can be used as a feedback signal for retard limit control. Engine MBT timing can also be derived from in-cylinder ionization signals similar to the pressure signals [63, 69]. Since MBT criteria derived from pressure and ionization signals are solely based upon observations and may change at different operating conditions, the associated control algorithms still require some dynamometer-based calibration effort. It is clear that the combustion process has to be matched with the engine cylinder volume change to attain the best torque. The major advantage for the ionization-based closed-loop MBT timing control is that no additional sensing element or assembly steps are required since it uses the spark plug as an ignition actuator and a combustion sensor.

This section proposes a closed-loop ignition control architecture (see Figure 2.24) that combines MBT timing control, knock, and retard timing limit control strategies into

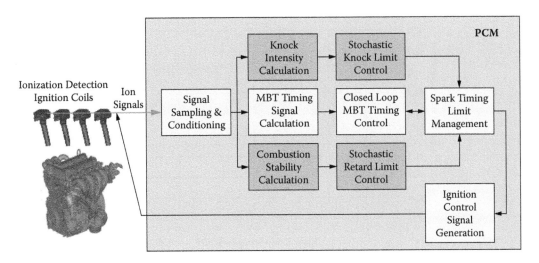

FIGURE 2.24
Closed-loop ignition timing control system.

an integrated one. The integrated ignition control architecture allows the engine to operate at its true MBT timing when it is not limited by borderline knock limit and operate at its borderline knock limit when it is limited by knock. Alternatively, it allows the engine to be operated at its borderline retard limit when it is limited by combustion stability.

2.5.1 Ignition System

The engine ignition system is designed to initiate the combustion process by igniting the air and fuel mixture trapped inside the cylinder. The main control parameters are the ignition timing and dwell duration, and normally the ignition starts at the end of dwell. Figure 2.25 shows a sample relationship between engine torque output and spark timing with all other engine control parameters fixed, and it can be found that there exists an optimal ignition timing that provides the highest torque. Therefore, the engine igniting needs to be optimized for the best fuel economy, which leads to closed-loop MBT timing control using either in-cylinder pressure or ionization signal. The MBT timing control using the in-cylinder pressure control is introduced in [65] and [68], and this section will concentrate on the closed-loop MBT timing control using ionization signal and mainly provides the MBT timing detection and other combustion information based upon the in-cylinder ionization signal [70, 71].

2.5.2 MBT Timing Detection and Its Closed-Loop Control

For the closed-loop MBT timing control strategy, an individual cylinder ionization current was sampled at every crank degree and every combustion event processed to generate both a composite MBT timing feedback criterion and closed-loop MBT timing control output. This section describes the MBT timing signal calculation and closed-loop MBT timing control blocks in Figure 2.24.

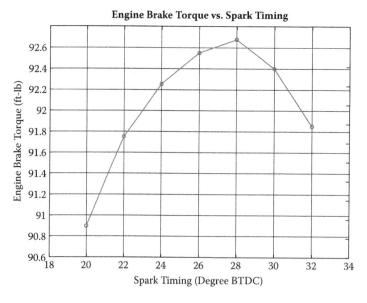

FIGURE 2.25
Engine torque and ignition timing.

2.5.2.1 Full-Range MBT Timing Detection

The MFB is determined by the well-known Rassweiler–Withrow [62] method, established in 1938, through pressure measurement. Through MFB one can find when the combustion has its peak burning velocity, acceleration, and percentage burn location as a function of crank angle. Maintaining these critical events at a specific crank angle produces the most efficient combustion process. In other words, the MBT timing can be found through these critical events. In reference [68], instead of directly using the MFB, the connection between MFB and net pressure is utilized to simplify analysis. The net pressure and its first and second derivatives are used to represent the distance, velocity, and acceleration of the combustion process. References [68] and [69] show that PCP location, 50% MFB location, and maximum acceleration location of the net pressure can each be used as MBT timing criteria for closed-loop control.

Figure 2.26 shows a typical ionization signal vs. crank angle and the corresponding in-cylinder pressure signal. Different from an in-cylinder pressure signal, an ionization signal actually shows more detailed information about the combustion process through its waveform. It shows when a flame kernel is formed and propagates away from the gap, when the combustion is accelerating rapidly, when the combustion reaches its peak burning rate, and when the combustion ends. A typical ionization signal usually consists of two peaks. The first peak of the ion signal represents the flame kernel growth and development, and the second peak is the reionization due to the in-cylinder temperature increase resulting from both pressure increase and flame development in the cylinder.

The use of an ionization signal for MBT timing detection was studied in [73]. As described in [73], the inflection point right after the first peak (called the first inflection point; see Figure 2.26) can be correlated to the maximum acceleration point of the net pressure, and this point is usually between 10% and 15% MFB. The inflection point right before

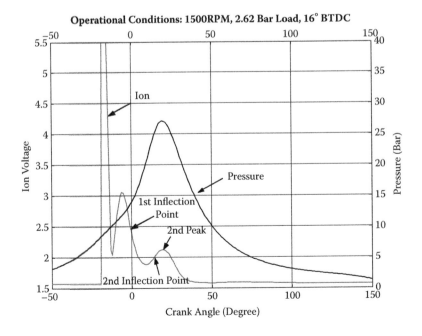

FIGURE 2.26
A typical ionization signal as function of crank angle.

the second peak of the ionization signal (called the second inflection point; see Figure 2.26) correlates well with the maximum heat release point and is located right around the 50% MFB location. Finally, the second peak location is related to the peak pressure location of the pressure signal (see Figure 2.26).

As described in [68, 69, 71], at MBT timing the maximum acceleration point of MFB (MAMFB) is located at around TDC, the 50% MFB location (50% MFB) is between 8° and 10° after TDC, and the peak cylinder pressure location (PCPL) is around 15° after TDC. Using the MBT timing criteria relationship between in-cylinder pressure and in-cylinder ionization signal, these three MBT timing criteria (MAMFB, 50% MFB, and PCPL) can be obtained using an in-cylinder ionization signal.

It is well known that the second peak of the ionization signal is mainly due to the high in-cylinder temperature during the combustion process. In the case that in-cylinder temperature cannot reach the reionization temperature threshold, the second peak of the ionization signal may disappear. For example, when the engine is operated either at the idle condition, with a very high EGR rate, or with a lean air-to-fuel (A/F) mixture, or a combination of the above, the flame temperature is relatively low and the temperature could be below the reionization temperature threshold. Therefore, the second peak may not appear in the ionization signal.

Previously mentioned MBT timing correlations over the entire operating range were presented in [68, 73] and are omitted here for brevity. During this study, it was observed that the following three cases cover all the possible ionization signals over the speed and load map:

Case 1: Normal ionization waveform. Both peaks are present in the waveform.

Case 2: The first peak ionization signal; low combustion temperature resulting in no second peak.

Case 3: The second peak ionization signal; high engine speed such that the first peak merges with the ignition signal due to the relatively longer ignition duration as a result of a relatively constant spark duration at high engine speed.

Figure 2.27 shows a representative example from each of the three cases. For the case 1 example, the engine was operated at 1500 rpm with 2.62 bar brake mean effective pressure (BMEP) load and without EGR; for the case 2 example, the engine was running at the same condition as case 1, except with 15% EGR; and for the case 3 example, the engine was running at 3500 rpm with wide-open throttle (WOT).

It is clear from Figure 2.27 that three MBT timing criteria (MAMFB, 50% MFB, and PCPL) are available only in case 1, and for cases 2 and 3, only one or two criteria are available. This indicates that at some operating conditions, only one or two MBT timing criteria can be obtained for MBT timing feedback. The proposed MBT timing estimation method combines all MBT timing criteria available at current operational conditions into one single composite criterion for improved reliability and robustness. The detailed algorithm is described in the next section.

In order to implement the MBT timing estimation strategy using an in-cylinder ionization signal, a detection algorithm was developed in [71]. The MBT timing detection algorithm can be divided into the following four steps:

Step 1: Ionization signal conditioning. For each given cylinder, the ionization signal is sampled at every crank degree after the ignition coil dwell event for 120°, as the ionization signal disappears after 120 crank angle degrees. The sampled

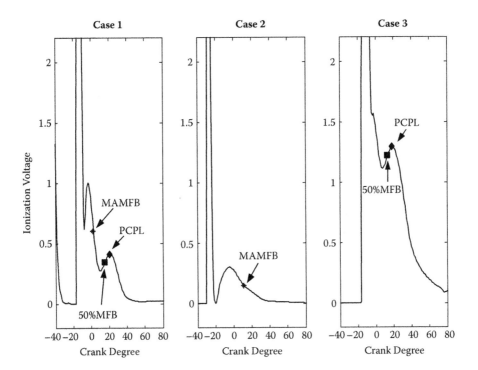

FIGURE 2.27
Three cases of ionization waveforms.

ionization signal *Ion*(·) is conditioned by a low-pass filtering to improve the accuracy of detecting the first and second peaks, and inflection points. In order to minimize the phase shift effects due to low-pass filtering for improved MBT timing estimation, a two-way low-pass filtering technique is used; see [68] for details. Note that the complete ionization signal array is available for computing the spark timing control for the next combustion event, so it is possible to perform this noncausal calculation. The ionization vector is filtered by the first-order forward filter defined below:

$$F_F(z) = \frac{1-a}{1-a \cdot z^{-1}}, \tag{2.50}$$

where *a* is the digital filter parameter associated with the low-pass filter bandwidth, and then the index of the ionization vector is reversed and filtered by $F_F(z)$ again. The combined filtering transfer function has zero phase delay; see [68].

Step 2: Operational condition identification. In this step, the engine operational condition is identified, and the resulting output of this step is the determination of which case the sampled ionization signal belongs to, that is, case 1, 2, or 3. A pattern recognition algorithm is used for the case identification by using the calculated number of peaks, inflection points, and their distances from the end of ignition.

Step 3: MBT timing criteria calculation. After the ionization signal case is identified, MBT timing criteria can be calculated using a peak location detection algorithm.

TABLE 2.3

Coefficient Selection Matrix

Coefficient	Case 1	Case 2	Case 3
α_{MAMFB}	1	1	0
$\alpha_{50\%MFB}$	1	0	1
α_{PCPL}	1	0	1

The inflection location detection logic is implemented by applying a peak location detection algorithm to the derivative of the filtered ionization signal.

Step 4: Composite MBT timing criterion generation. The composite MBT timing criterion C_{MBT} is calculated based upon the case number identified from step 2. For all three cases, the composite MBT timing criterion can be calculated using the following equation:

$$C_{MBT} = \frac{[\alpha_{PCPL} \cdot (PCPL - PCPL_{OFFSET}) + \alpha_{50\%MFB} \cdot (50\%MFB - 50\%MFB_{OFFSET}) + \alpha_{MAMFB} \cdot MAMFB]}{\beta}$$

(2.51)

where, $\beta = \alpha_{MAMFB} + \alpha_{50\%MFB} + \alpha_{PCPL} \neq 0$.

By definition, the composite MBT timing criterion is equal to zero when the engine is running at its MBT timing condition since the MBT timing criterion MAMFB is zero and the MBT criteria 50% MFB and PCPL are shifted from their nominal locations defined by $50\%MFB_{OFFSET}$ and $PCPL_{OFFSET}$, respectively. At MBT timing both $50\%MFB_{OFFSET}$ and $PCPL_{OFFSET}$ vary slightly (a few degrees) as a function of engine operational conditions, and they are a function of engine speed and load. They can be obtained through an existing calibration process (no extra calibration needed). Coefficients α_{MAMFB}, $\alpha_{50\%MFB}$, and α_{PCPL} are selected based upon Table 2.3 for a given case number.

2.5.2.2 Closed-Loop MBT Timing Control

The purpose of closed-loop MBT timing control is twofold: keeping the engine operating at its MBT spark timing if it is not knock limited and reducing the cycle-to-cycle combustion variation through closed-loop spark timing control [68]. The control strategy associated with the MBT timing signal calculation and closed-loop MBT timing control blocks, shown in Figure 2.24, is discussed below.

Inputs to the MBT timing signal calculation block (see Figure 2.24) are the individual in-cylinder ionization signals synchronized with engine crank angle and the current engine operational information, such as the engine speed, load, etc. Speed and load information are lookup table inputs for calculating MBT timing offsets ($PCPL_{OFFSET}$ and $50\%MFB_{OFFSET}$), as well as the MBT feedforward ignition timing. The output of the MBT timing signal calculation block is the composite MBT timing criterion C_{MBT} for the current cylinder using the proposed algorithm from the previous subsection.

The closed-loop spark MBT timing control is realized using a PI controller whose output is used to correct the feedforward MBT ignition timing. The error between the MBT criterion reference and the composite MBT timing criterion is used as an input to the PI controller. Under the study conditions used in this work, the MBT criteria reference signal was set to zero. However, this is not necessary. For example, under certain conditions, to meet emission requirements it may be desired to retard spark timing from its MBT timing by a few degrees. Under these conditions, the reference signal can be set to a negative value.

2.5.3 Stochastic Ignition Limit Estimation and Control

This section describes a stochastic ignition limit control algorithm that is used for both advanced and retard ignition timing limit controls.

2.5.3.1 Stochastic Ignition Limit Estimation

This subsection discusses how to generate a feedback measure for the retard limit control, which is the functionality of the combustion stability calculation block shown in Figure 2.24. A typical ionization signal with its integration window is displayed in Figure 2.28. Before defining the combustion stability criterion, let us define an integral ratio function $R_{INT}(\cdot)$ as follows:

$$R_{INT}(k) = \frac{\left(\displaystyle\sum_{i=CS}^{CS+k-1} Ion(i) \right)}{\left(\displaystyle\sum_{i=CS}^{CS+n-1} Ion(i) \right)} \times 100\%, \quad k \le n \tag{2.52}$$

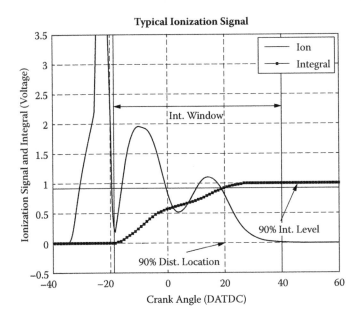

FIGURE 2.28
Ionization signal and integration location.

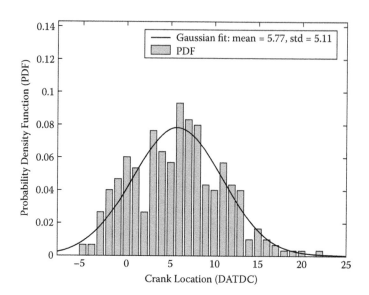

FIGURE 2.29
Ion integration location PDF.

where $Ion(\cdot)$ is a 120-element ionization vector sampled at the start of ignition with one crank degree resolution (same vector for the MBT timing estimation), CS is the crank angle index at the start of the integration window (see Figure 2.28), and n is the crank degrees representing the integration window width. The integration location (IL) of a given percentage R_{DES} is an integer $IL(R_{DES})$ that satisfies the following equation:

$$R_{INT}[IL(R_{DES}) - 1] < R_{DES} \leq R_{INT}[IL(R_{DES})] \qquad (2.53)$$

Figure 2.28 shows a 90% integration location $IL(90\%)$. Note that 100% integration location $IL(100\%)$ is ideally reached at the end of the integration window.

Figure 2.29 shows the stochastic properties of $IL(90\%)$ with spark timing at 21° before TDC. Three hundred cycles (number of consecutive firing events at the same spark timing) of data are used to create the probability density function (PDF) or histogram of the integration location, where the solid line is a Gaussian distribution fit of PDF data.

Based on the PDF shown in Figure 2.29, the statistics of the ionization integration locations seem to be close to those of a Gaussian random process. As the spark timing becomes more retarded, the PDF of integration location starts skewing toward the retard direction (see [77] for details). But more importantly, at the spark timing with a desired combustion stability level, the PDFs are close to those of a Gaussian random process.

The mean and standard deviation of the ionization integration locations (90%) during a spark sweep at 1500 rpm with 2.62 bar BMEP are shown in Figure 2.30, where stars represent the test data and the solid lines are fitted curves using polynomials. It can be observed that both mean and standard deviation of integration location increase as the spark timing retards.

2.5.3.2 Knock Intensity Calculation and Its Stochastic Properties

This subsection is associated with the knock intensity calculation block in Figure 2.24. Under engine operational conditions that result in knock, knock intensity can be calculated

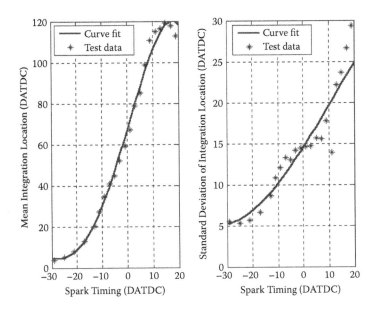

FIGURE 2.30
Mean and standard deviation of integration locations.

using in-cylinder ionization signals in a similar way to using an in-cylinder pressure sensor signal [65, 74, 76]. Knock intensity calculation utilizes the high-frequency component (between 3 and 15 kHz corresponding to the knock frequency range) of the ionization signal over a given knock window defined in Figure 2.28. An analog circuit was used to calculate knock intensity. Define $Ion_{KNK}(\cdot)$ as the analog ionization signal and $Ion_{KNK-BP}(\cdot)$ as the band-pass-filtered ionization knock signal that is obtained by filtering $Ion_{KNK}(\cdot)$ using a fourth-order Butterworth band-pass filter with corner frequencies of 3 and 15 kHz, respectively. The knock intensity I_{KNK} can be calculated using the following formula:

$$I_{KNK} = \int_{T_1}^{T_2} |Ion_{KNK-BP}(t)| dt \qquad (2.54)$$

where T_1 is the time corresponding to the beginning of the knock window defined in Figure 2.28, and T_2 is the time associated with the end of the knock window. Figure 2.31 shows a PDF of knock intensity signal obtained from the ionization signal, where an analog circuit was used for calculating knock intensity I_{KNK}. The engine was operated at 1000 rpm with WOT. The spark timing is at 18° before TDC. Comparing the PDF drawings of Figures 2.29 and 2.31, the knock intensity PDF histogram is not symmetric, and it is obvious that a Gaussian random process cannot approximate it. In fact, reference [81] shows that the knock intensity I_{KNK} is a lognormal random process.

The mean and standard deviation of knock intensity I_{KNK} during a spark sweep at 1500 rpm with 2.62 bar BMEP are shown in Figure 2.32, where stars represent the test data and the solid lines are fitted curves using polynomials. It can be observed that both the mean and standard deviation of the integration location decrease as the spark timing retards. For stochastic limit control we use the mean of knock intensity and the information contained in the probability density function, such as percentage of knock intensity below a given threshold, as the feedback signals.

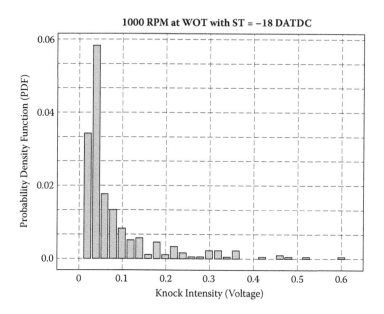

FIGURE 2.31
Knock intensity PDF.

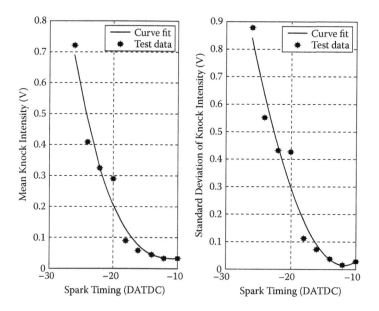

FIGURE 2.32
Mean and standard deviation of knock intensity.

2.5.3.3 Stochastic Limit Control

The objective of the stochastic limit controller is to provide dynamic ignition timing limits for the overall spark controller to avoid engine knock in the advanced direction or to assure combustion stability in the retard direction. This subsection describes the strategies of both the stochastic knock limit control and the stochastic retard limit control blocks in

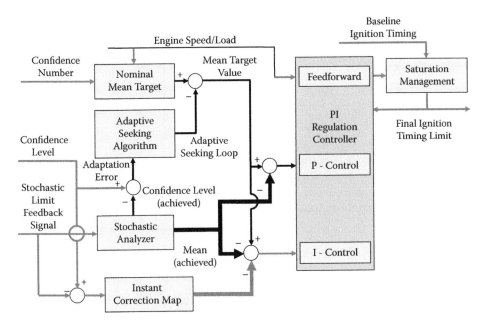

FIGURE 2.33
Stochastic closed-loop retard limit controller.

Figure 2.24. Since the stochastic knock limit controller structure is the same as the retard limit control, it will not be discussed in detail.

Figure 2.33 shows the architecture of the closed-loop stochastic limit controller for the retard limit application. Inputs to this controller block are user-specified confidence level targets, made up of two parts (reference confidence number CN_{REF} and level CL_{REF}), and the stochastic limit feedback signal $IL(R_{DES})$. The control objective is to maintain a reference confidence number (percentage) CN_{REF} of the controlled feedback signal $IL(R_{DES})$ below the reference confidence level CL_{REF}.

There are three main feedback actions of the proposed control scheme shown in Figure 2.33. Their functionalities are discussed below.

2.5.3.3.1 Adaptive Seeking Feedback (Thin Black Lines)

The purpose of this loop is twofold: reducing the calibration conservativeness by providing the regulation set point with its "true" mean target value and improving robustness of the stochastic limit controller when the engine operates under different conditions. This control loop is associated with two blocks in Figure 2.33. They are the stochastic analyzer and the adaptive seeking algorithm blocks. The nominal mean target block consists of a multidimensional lookup table using reference confidence number CN_{REF}, engine speed, and load as input, and the output is the estimated mean target MT from a calibration table. MT is the desired value for the mean of the feedback signal. The stochastic analyzer block forms a buffer B_{IL} of $IL(R_{DES})$ with a calibratable length m (number of consecutive combustion events). At each event, the oldest date is replaced by the new one in the buffer. The mean of B_{IL} is calculated by

$$MN_{IL} = \frac{1}{m}\sum_{i=0}^{m-1} B_{IL}(i), \qquad (2.55)$$

and actual confidence number CN_{ACT} can be calculated by

$$CN_{ACT} = \frac{\left(\sum_{i=0}^{m-1} B_{IL}(i) \cdot I_B(i)\right)}{\left(\sum_{i=0}^{m-1} B_{IL}(i)\right)} \times 100\%, \tag{2.56}$$

where $I_B(i) = 1$ if $B_{IL}(i) \leq CL_{REF}$; otherwise, $I_B(i) = 0$.

The actual confidence level CL_{ACT} of a given confidence number CN_{REF} is another parameter of interest. Define $\bar{B}_{IL}$ as a reordered vector of B_{IL} with its elements arranged in an increased order. Then the actual confidence level can be defined as follows:

$$CL_{ACT} = \bar{B}_{IL}(k). \tag{2.57}$$

where k is the closest integer of $m \cdot CL_{REF}$. The adaptive seeking algorithm utilizes adaptation error ($CL_{REF} - CL_{ACT}$) as input, and the output is mean target correction (MTC) obtained by integrating the adaptation error with a calibration gain. This control loop is used to reduce the conservativeness of the mean target MT for the regulation controller discussed next.

2.5.3.3.2 Regulation Stochastic Feedback (Thick Black Line)

The regulation loop is used to regulate the mean value of the stochastic limit feedback signal to a mean target value. The regulation controller is structured as a PI controller and a feedforward term based on engine operating conditions. The error input to the PI controller is

$$err_{PI} = MT - MTC - MN_{IL}. \tag{2.58}$$

Despite the variability of the stochastic retard limit feedback signal $IL(R_{DES})$, its mean value is a well-behaved signal for regulation purposes. The regulation controller is tuned to provide the desired settling time and steady-state accuracy for the response.

2.5.3.3.3 Instant Correction Feedback (Thick Gray Line)

This block calculates an instant correction signal to be fed into the integration portion of the PI controller. The instant correction is generated by a lookup table using the error signal $CL_{REF} - IL(R_{DES})$ as input. When the error is greater than zero, the output is zero, and when the error is less than zero, the output is positive and increases as the input becomes more negative. Feeding the instant correction to the integral term is equivalent to the role of counter-up/down logic. Instant correction refers to a spark retard action at the next combustion event; the gain determines how much spark retard is generated for the excessive knock intensity value over the threshold.

For knock limit control, the same controller structure was used. The stochastic limit feedback signal $IL(R_{DES})$ is replaced by calculated I_{KNK}, and the details can be found in [76].

The interaction of the stochastic limit controllers with the MBT controller is managed by the spark timing limit management block in Figure 2.24. This interaction may be illustrated using the retard limit control as an example. If the baseline spark timing is more advanced than the current retard limit, then the baseline spark is used as it is. In that case, the retard limit controller pushes the limit in the maximum retard direction by itself. This is due to the fact that the limit controller integrator has a negative input and keeps integrating until the maximum retard allowed is reached (an antiwindup scheme is used). If the baseline spark controller pushes the spark timing to a level at which the feedback signals generate corrections, then the retard spark limit moves from its maximum retard spark limit to a less retarded level as a variable saturation limit on the baseline spark. On the other hand, if the baseline spark continues pushing the spark in the retard direction even when the baseline spark is saturated, the seeking and instant correction actions of the retard controller will adjust the retard limit online.

2.5.4 Experimental Study Results

The experimental study of the proposed spark timing control strategy consists of three subsections. These subsections present the experimental results for closed-loop MBT timing and advanced and retard limit controllers. At the end of this section the results of an integrated ignition control system that combines the MBT timing controller with both advanced and retard limit controllers are shown.

The proposed control system was validated in an engine dynamometer. The engine was controlled by the engine dynamometer except for engine spark timing. The engine dyna-mometer controlled the engine throttle position, EGR rate, and fuel injection. It also con-trolled the engine speed and load. A rapid prototype controller was used for prototyping both open-loop and closed-loop spark timing control. Laboratory-grade pressure sensors were installed in every cylinder for comparing with in-cylinder ionization signals.

The digital waveform capture card inside the rapid prototype controller generates the interrupt based upon the crankshaft encoder pulses that trigger the data sampling process. A calculated crank angle also generates a software interrupt to initiate the combustion event-based closed-loop ignition control calculation. The spark timing is calculated during exhaust stroke of the corresponding cylinder to make the spark control commands avail-able before the intake stroke. The closed-loop control algorithms, shown in Figure 2.33, run every combustion event.

2.5.4.1 Closed-Loop MBT Timing Control

Before the results of closed-loop MBT timing control are discussed, results demonstrating the validity of the composite MBT timing criterion are presented on an example operating point. Figure 2.34 shows the estimated MBT timing criterion C_{MBT} when the engine was running at 1500 rpm with 7.0 bar BMEP load, and the ignition timing is at its MBT timing (21° before TDC). The data shown in Figure 2.34 are 100 cycles of data for cylinder 3 only. It can be observed that the MBT criterion confirms that the engine is operated close to its MBT timing since the average of the calculated C_{MBT} is close to zero. This ionization MBT timing detection algorithm was validated over the entire engine operational map using offline test data.

Figure 2.35 shows the relationship between estimated ionization MBT criterion C_{MBT} and in-cylinder pressure MBT criteria MAMFB, 50% MFB, and PCPL, where MAMFB is the maximum acceleration point of MFB calculated from in-cylinder pressure signal [68], which is defined as the peak location of the second derivative of MFB. The data shown in Figure 2.35 are a result of spark timing sweep when the engine was operated at 1500 rpm with 7.0 bar BMEP load. The spark timing varies from 13° to 25° before TDC. The top graph of Figure 2.35 shows the MBT timing criteria for cylinder 3, and the bottom one shows

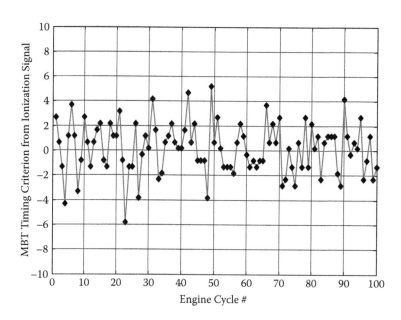

FIGURE 2.34
MBT criterion from ionization signal.

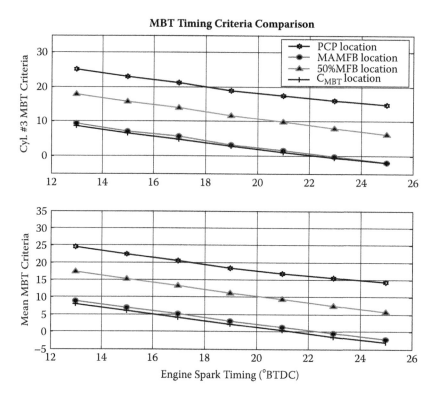

FIGURE 2.35
MBT timing criteria vs. timing sweep.

the average MBT timing criteria over all the cylinders. It is clear that the ionization MBT timing criterion C_{MBT} is very close to the MAMFB.

From the average MBT timing criteria plot, it can be concluded that the engine MBT timing is around 21° before TDC since both C_{MBT} and MAMFB are close to zero at that timing. An important observation of both top and bottom graphs is that all four MBT timing criteria are almost linear against the spark timing sweep, which provides for good closed-loop control characteristics. Test data for the other cylinders are also similar to those for cylinder 3.

Figure 2.36 shows the response of the PCP location, 50% MFB location, and C_{MBT} under closed-loop MBT timing control at 1500 rpm with 7.0 bar BMEP load. It is clear that all three MBT timing criteria on average remain at their MBT locations, respectively. That is, the PCP location is around 14° to 16° after TDC, the 50% MFB location is around 8° to 10° after TDC, and C_{MBT} is close to TDC. The closed-loop controller demonstrated in this chapter utilizes ionization signals from all the cylinders continuously to generate a global ignition timing control signal. Constant timing offsets are then used to compensate for individual cylinder unbalance. The PI gains are tuned to have a sufficient stability margin.

Another aspect of analyzing closed-loop control of engine MBT timing is from a stochastic perspective. It is well known that for a linear dynamic system with a stationary stochastic process input, closed-loop controllers, such as a linear quadratic Gaussian (LQG) controller, are able to reduce closed-loop system output variances. Figure 2.37 shows output variances of MBT timing criteria (PCP and 50% MFB locations) for both open-loop and closed-loop control. The engine operating condition is the same as the open-loop control case shown in Figure 2.34. It is clear that closed-loop control using ionization-based MBT timing feedback reduces cycle-to-cycle variances shown in Figure 2.37 by about 5% to 10%, resulting in a smoother-running engine.

Finally, Figure 2.38 shows the results of closed-loop ionization MBT timing control during the engine warm-up process. The engine was operating at 1500 rpm with 64 Nm (Newton-meter) load. The whole process lasted for about 10 min and resulted in almost

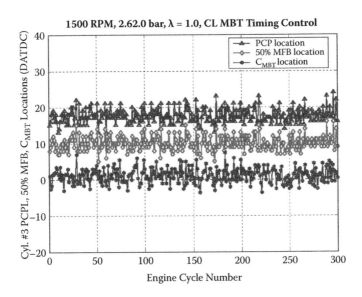

FIGURE 2.36
CL MBT timing control with ionization feedback.

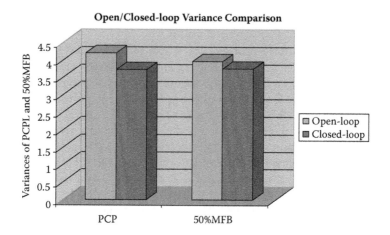

FIGURE 2.37
Open/closed-loop variance comparison.

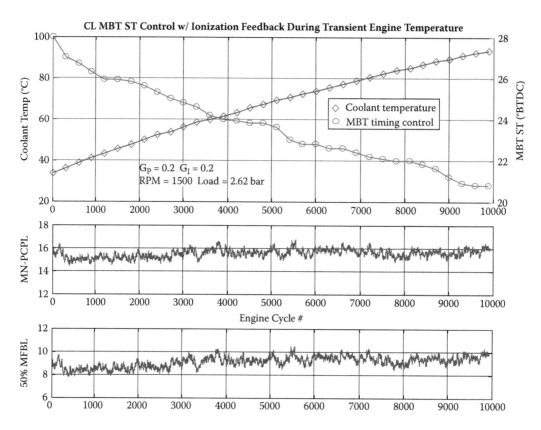

FIGURE 2.38
CL control for transient temperature.

10,000 engine cycles. The starting engine coolant temperature is about 34°C, and the ending temperature is about 93°C. The top graph of Figure 2.38 shows the relationship between engine coolant temperature and engine MBT timing. In order to keep the composite ionization MBT timing criterion at or around the TDC location, the ignition timing has to be advanced to compensate for relatively slow combustion when the engine is cold. It is clear that during the 10 min warm-up process, the closed-loop MBT timing controller moves the spark timing in a retard direction from around 28° before TDC to 21° before TDC. During this warm-up process, the burn rate increases and the corresponding MBT spark timing moves back in a retard direction.

The second graph from the top in Figure 2.38 shows the average PCP location of all cylinders, and the bottom graph shows the average 50% MFB locations for all cylinders. It can be observed that the mean PCP location is between 15° and 16° after TDC during the engine warm-up process, and that the average 50% MFB location for all four cylinders is between 8° and 10° after TDC. This validates that the engine operates at its MBT timing during the engine warm-up process, and it also shows that closed-loop MBT timing control using ionization feedback is able to operate the engine at its MBT timing during the temperature transition.

2.5.4.2 Closed-Loop Retard Limit Control

During an engine cold-start process, the stochastic retard limit manager seeks the maximum retard possible while assuring that misfire and partial burn are avoided with the objective of increasing the catalyst temperature rapidly to minimize tailpipe emissions. Delaying the combustion through high values of ignition retard can shorten the time that it takes the catalyst to reach its light-off temperature. Therefore, the conventional three-way catalyst becomes effective much sooner in reducing tailpipe emissions [78–80]. However, if the ignition retard is too much, engine-out HC emissions become excessive due to incomplete combustion (partial burn) as well as misfire. An open-loop retard calibration needs to provide enough margins to avoid misfire under all conditions and with all fuels. It therefore is inherently conservative.

The calibration of the stochastic retard limit controller (see Figure 2.33) for cold-start retard limit control can be explained as follows: suppose that we want to make sure that a given percentage CN_{REF} of the integration locations will not go beyond a certain crank angle (say, 110° after TDC). Since the integration location practically represents the end of the combustion, this is equivalent to saying that for a CN_{REF} percentage of the consecutive combustion events the combustion will be over before the desired crank angle (110° after TDC in this example). This location is then the desired confidence level target CL_{REF} for the feedback control. In this sense, CN_{REF} represents the acceptable combustion retard in crank degrees, which will be continuously monitored from the ion current processing. Using the standard deviation of the measured data, a nominal target mean for the regulation controller is calculated by subtracting a certain multiple of the standard deviation of the measured data in the buffer. That initial mean target is then increased by the adaptive seeking loop slowly if the actual confidence level (say, with 90% confidence number) $CL_{ACT}(CN_{REF})$ computed from the measured data is less than the desired confidence level of 110° after TDC.

Figures 2.39 and 2.40 show responses from a cold-start run using integration location as the feedback signal. The 90% confidence level, $CL_{ACT}(90\%)$, is also included in Figure 2.39 as a performance measure. Note that it was kept around 110° after TDC at the steady state and did not exceed 124° after TDC during the transient operation, which was the exhaust

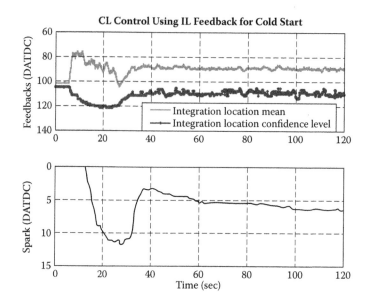

FIGURE 2.39
Control for cold-start run-up.

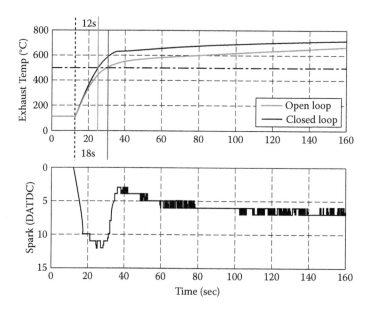

FIGURE 2.40
Temperature vs. controlled ignition retard.

valve opening timing for the particular engine tested. Therefore, combustion was completed before the exhaust valves were opened, which is critical for engine HC emissions.

Figure 2.40 demonstrates the corresponding exhaust temperature rise during the run. An open-loop temperature profile was also included in Figure 2.40 to show the improved temperature rise time with the proposed control. For the open-loop case, the ignition timing was held at TDC, which was the initial ignition timing for the closed-loop controller.

Based on Figure 2.40, the time it takes the exhaust temperature to reach 500° C was reduced from 18 s to 12 s using the closed-loop controller.

2.5.4.3 Closed-Loop Knock Limit Control

Before applying stochastic limit control to knock limit management, knock controllability was studied using ionization knock intensity feedback. Due to the high-resolution knock intensity signal obtained from in-cylinder ionization signals, both mean and standard deviation of the knock intensity signal show high correlation to engine spark timing (see Figure 2.41).

Both mean and standard deviation of knock intensity increase when the engine spark timing varies from 10° before TDC to 26° before TDC. This demonstrates good controllability using the knock intensity obtained from ionization signals. The mean and standard deviation data are processed using 300-cycle ionization data. Similar results are obtained over the whole speed and load range of the engine.

The knock intensity actual confidence levels $CL_{ACT}(90\%)$, $CL_{ACT}(95\%)$, and $CL_{ACT}(100\%)$ are shown in Figure 2.42. The actual confidence levels of 90%, 95%, and 100% increase as the spark timing advances. The criterion used for adaptive seeking is the actual confidence level $CL_{ACT}(CN_{REF})$ with reference confidence number for stochastic limit control. This adaptive seeking control loop adjusts the reference signal for the mean control loop such that the given confidence number percentage CN_{REF} of the actual knock intensity signal stays below the target confidence level CL_{REF}.

The closed-loop control results of the knock intensity confidence number and level, using the proposed stochastic limit control of Figure 2.33, are shown in Figure 2.43, where the top plot shows both actual mean knock intensity and actual confidence level of knock intensity with reference confidence number 90%, the second plot from the top shows the spark advanced limit and actual spark timing, the third shows the instantaneous knock

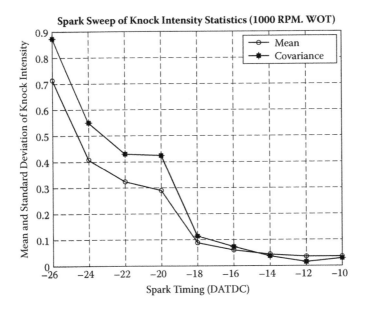

FIGURE 2.41
Knock intensity statistics.

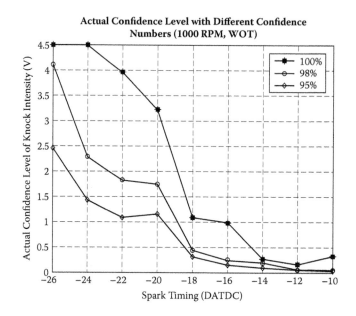

FIGURE 2.42
Knock intensity actual confidence levels.

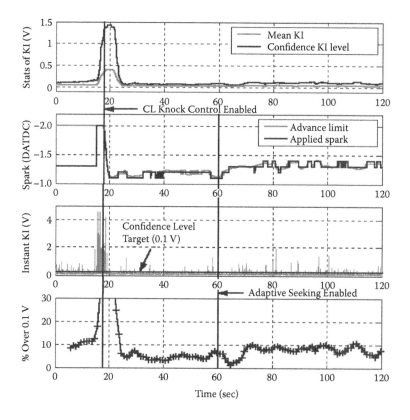

FIGURE 2.43
CL knock limit control.

intensity (KI) signal and the reference confidence level knock intensity at 0.1 V, and the bottom plot shows the percentage of KI over the 0.1 V threshold ($100\% - CN_{ACT}(0.1V)$). Note that the control calibrations (CN_{REF} is 90% and CL_{REF} for KI is 0.1 V) set the knock limit control objective as keeping 90% of the consecutive knock intensity levels below 0.1 V. During the first 18 s, the closed-loop knock limit control is not active, baseline spark timing starts at around 13° before TDC, the advanced limit is at its maximum of 20° before TDC, and the knock intensity mean is relatively low (less than 0.1 V). At 16 s, the baseline spark timing is manually advanced to 20° before TDC and mean and actual confidence level knock intensity increases to over 0.40 and 1.4 V, respectively, right before the closed-loop knock limit control is activated. After the mean knock limit control is enabled at 18 s, knock intensity is reduced to the desired knock intensity level and the advanced limit, generated by the closed-loop knock limit controller, moves to the 11~12° range before TDC. Note that the advanced spark is digitized from the advanced limit with 1° resolution due to the control hardware limitations. Between 18 and 60 s, the bottom plot shows that there is about 5% of the actual knock intensity staying beyond the 0.1 V reference confidence level. At 60 s, the adaptive seeking algorithm is enabled with a 90% reference confidence number, and the KI percentage over 0.1 V target increases to around 10% (or equivalent to 90% actual confidence level). The spark timing is further advanced to between 14° and 13° before TDC.

Figure 2.44 shows the test results of the combined advanced (knock) and retard limit control. The thin dark line is the engine baseline spark timing starting at 15° before TDC. Since the engine is neither knock limited nor retard limited, both the advanced (thick gray) and retard (thin gray) limits stay at their maximum levels (40° before TDC for advanced limit and 5° before TDC for retard limit). When the baseline spark timing moves in the advanced direction and causes engine knocking, the advanced limit reduces due to the closed-loop knock limit control, the baseline spark timing is limited to about 23° before TDC, and the retard spark limit still stays at its maximum retard limit (5° before TDC). At 30 s, the baseline spark timing is manually moved in the retard direction. At about 38 s, the retard limit control moves the retard limit in the advanced direction due to the reduced

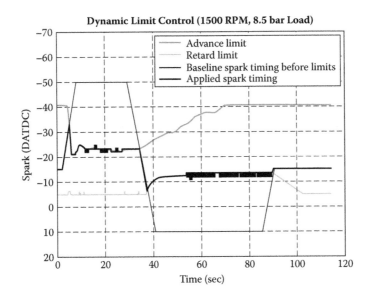

FIGURE 2.44
CL advanced and retard limit control.

combustion stability and the retard spark timing stabilizes at about 12° before TDC, while the knock limit controller independently returns the advanced limit to its maximum at 40° before TDC due to its integral action since knock is below the target. This plot demonstrates both steady-state and transitional control utilizing both knock advanced limit control and combustion stability retard limit control. It also shows how each limit control interacts with the baseline (or MBT) ignition timing control as an independent timing limit in both directions.

As a summary, the closed-loop ignition control architecture discussed in this section combines three closed-loop ignition control strategies into a single one. They are closed-loop MBT timing control, borderline knock limit control, and retard limit control. The integrated ignition control architecture allows the engine to operate at its true MBT timing when it is not limited by borderline knock limit and operate at its borderline knock limit when it is knock limited. During a cold start, the closed-loop controller operates the engine at its maximum retard limit for fast catalyst light-off while maintaining combustion stability at the desired level. The control strategy has been validated under steady-state and slow transient operations and the fast transient tests (such as the Federal Test Procedure (FTP)) have not been completed yet. Combining this quasi-steady-state-oriented ignition timing control strategy with a feedforward controller for improved transient performance remains a work in progress.

References

1. M. Jung and R.G. Ford, Parameterization and Transient Validation of a Variable Geometry Turbocharger for Mean Value Modeling at Low and Medium Speed-Load Points, SAE Technical Paper 2002-01-2729, 2002.
2. M. Canova, R. Garcin, M.M. Shawn, Y. Guezennec, and G. Rizzoni, A Control-Oriented Model of Combustion Process in a HCCI Diesel Engine, presented at Proceedings of the American Control Conference, Portland, OR, June 8–10, 2005.
3. G.G. Zhu, I. Haskara, and J. Winkelman, Closed-Loop Ignition Timing Control for SI Engines Using Ionization Current Feedback, *IEEE Transactions on Control System Technology*, 15(3), 2007.
4. J. Bengtsson, P. Strandh, R. Johansson, P. Tunest, and B. Johansson, Closed-Loop Combustion Control of Homogeneous Charge Compression Ignition (HCCI) Engine Dynamics, *International Journal of Adaptive Control and Signal Processing*, 18: 167–179, 2004.
5. F. Willems, E. Doosje, F. Engels, and X. Seykens, Cylinder Pressure-Based Control in Heavy-Duty EGR Diesel Engines Using a Virtual Heat Release and Emission Sensor, SAE Technical Paper 2010-01-0564, 2010.
6. M.M. Andreae, W.K. Cheng, T. Kenney, and J. Yang, On HCCI Engine Knock, SAE Technical Paper 2007-01-1858, 2007.
7. S.M. Aceves, D.L. Flowers, R.W. Dibble, and A. Babajimopoulos, Overview of Modeling Techniques and Their Application to HCCI/CAI Engines, in *HCCI and CAI Engines for the Automotive Industry*, ed. H. Zhao, Cambridge: Woodhead Publishing, 2007.
8. O. Colin, A.P. Cruz, and S. Jay, Detailed Chemistry-Based Auto-Ignition Model Including Low Temperature Phenomena Applied to 3-D Engine Calculations, *Proceedings of the Combustion Institute*, 30: 2649–2656, 2005.
9. P.M. Diaz, D. Prasad, and S.M. Raman, A CFD Investigation of Emissions Formation in HCCI Engines, Including Detailed NOx Chemistry, *IETECH Journal of Mechanical Design*, 1(1): 056–061, 2007.

10. Gamma Technology, *Engine Performance Application Manual*, GT-Suite version 7.0, September 2009. Available at http://www.gtisoft.com/.
11. L. Guzzella and C.H. Onder, *Introduction to Modeling and Control of Internal Combustion Engine Systems*, Berlin: Springer, 2004.
12. S. Saulnier and S. Guiliain, Computational Study of Diesel Engine Downsizing Using Two-Stage Turbocharging, SAE Technical Paper 2004-01-0929, 2004.
13. A.Y. Karnik, J.H. Buckland, and J.S. Freudenberg, Electronic Throttle and Wastegate Control for Turbocharged Gasoline Engines, in *Proceedings of the American Control Conference*, Portland, OR, 2005, pp. 4434–4439.
14. K.L. Hoag and R.J. Primus, *Fundamentals of Diesel Engine Performance*, Cummins Engine Company, Columbus, IN, 1992.
15. X. Yang and G. Zhu, A Mixed Mean-Value and Event-Based Model of a Dual-Stage Turbocharged SI Engine for Hardware-in-the-Loop Simulation, presented at Proceedings of the American Control Conference, Baltimore, MD, June 2010.
16. X. Yang and G. Zhu, A Control Oriented Hybrid Combustion Model of an HCCI Capable SI Engine, *Journal of Automobile Engineering*, 226(10): 1380–1395, 2012.
17. S. Zhang, G. Zhu, and Z. Sun, A Control-Oriented Charge Mixing and Two-Zone HCCI Combustion Model, *IEEE Transactions on Vehicular Technology*, 63(3): 1079–1090, 2014.
18. M. Locatelli, C.H. Onder, and H.P. Geering, An Easily Tunable Wall-Wetting Model for PFI Engines, SAE Technical Paper 2004-01-1461, 2004.
19. J.B. Heywood, *Internal Combustion Engine Fundamentals*, McGraw-Hill, New York City, NY, 1988.
20. A. Lefebvre and S. Guilain, Modeling and Measurement of the Transient Response of a Turbocharged SI Engine, SAE Technical Paper 2005-01-0691, 2005.
21. G.M. Shaver, M.J. Roelle, and J.C. Gerdes, Modeling Cycle-to-Cycle Dynamics and Mode Transition in HCCI Engines with Variable Valve Actuation, *Control Engineering Practice*, 14: 213–222, 2006.
22. J. Chang and O. Guralp, New Heat Transfer Correlation for an HCCI Engine Derived from Measurements of Instantaneous Surface Heat Flux, SAE Technical Paper 2004-01-2996, 2004.
23. K. Inoue, K. Nagakiro, Y. Ajiki, and N. Kishi, A High Power Wide Torque Range Efficient Engine with a Newly Developed Variable Valve Lift and Timing Mechanism, SAE Technical Paper 890675, 1989.
24. Y. Moriya, A. Watanabe, H. Uda, H. Kawamura, M. Yoshioka, and M. Adachi, A Newly Developed Intelligent Variable Valve Timing System—Continuously Controlled Cam Phasing as Applied to a New 3 Liter Inline 6 Engine, SAE Technical Paper 960579, 1996.
25. K.H. Oehling, R. Teichmann, and H. Unger, Requirements and Potentials of Future Valve Train Concépts, SAE Technical Paper 964211, 1996.
26. R. Flierl and M. Kluting, The Third Generation of Valvetrains—New Fully Variable Valvetrain for Throttle Free Load Control, SAE Technical Paper 2000-01-1227, 2000.
27. M. Theobald, B. Lequesne, and R. Henry, Control of Engine Load via Electromagnetic Valve Actuators, SAE Technical Paper 940816, 1994.
28. P. Kreuter, P. Heuser, and M. Schebitz, Strategies to Improve SI-Engine Performance by Means of Variable Intake Lift, Timing and Duration, SAE Technical Paper 920449, 1992.
29. M. Schecter and M. Levin, Camless Engine, SAE Technical Paper 960581, 1996.
30. C. Turner, G. Babbitt, C. Balton, M. Raimao, and D. Giordano, Design and Control of a Two-Stage Electro-Hydraulic Valve Actuation System, SAE Technical Paper 2004-01-1265, 2004.
31. J. Allen and D. Law, Production Electro-Hydraulic Variable Valve-Train for a New Generation of I.C. Engines, SAE Technical Paper 2002-01-1109, 2002.
32. D. Denger and K. Mischker, The Electro-Hydraulic Valvetrain System EHVS—System and Potential, SAE Technical Paper 2005-01-0774, 2005.
33. J. Watson and R. Wakeman, Simulation of a Pneumatic Valve Actuation System for Internal Combustion Engine, SAE Technical Paper 2005-01-0771, 2005.
34. J. Ma, H. Schock, U. Carlson, A. Hoglund, and M. Hedman, Analysis and Modeling of an Electronically Controlled Pneumatic Hydraulic Valve for an Automotive Engine, SAE Technical Paper 2006-01-0042, 2006.

35. R.M. Richman and W.C. Reynolds, A Computer-Controlled Poppet-Valve Actuation System for Application on Research Engines, SAE Technical Paper 840340, 1984.
36. J. Turner, M. Bassett, R. Pearson, G. Pitcher, and K. Douglas, New Operating Strategies Afforded by Fully Variable Valve Trains, SAE Technical Paper 2004-01-1386, 2004.
37. Z. Sun and D. Cleary, Dynamics and Control of an Electro-Hydraulic Fully Flexible Valve Actuation System, in *Proceedings of the American Control Conference*, Denver, CO, 2003, pp. 3119–3124.
38. Z. Sun and T. Kuo, Electro-Hydraulic Fully Flexible Valve Actuation System for Advanced Combustion Development, in *Proceedings of the FISITA World Congress*, Yokohama, Japan, F2006P287, 2006.
39. Z. Sun and H. Xin, Development and Control of Electro-Hydraulic Fully Flexible Valve Actuation System for Diesel Combustion Research, SAE Technical Paper 2007-01-4021, 2007.
40. Z. Sun, Engine Valve Actuation Assembly with Dual Automatic Regulation, U.S. Patent 6,959,673, 2005.
41. P. Gillella and Z. Sun, Design, Modeling, and Control of a Camless Valve Actuation System with Internal Feedback, *IEEE/ASME Transactions on Mechatronics*, 16(3): 527–539, 2011.
42. Z. Sun, Electrohydraulic Fully Flexible Valve Actuation System with Internal Feedback, *Journal of Dynamic Systems, Measurement, and Control*, 131: 024502, 2009.
43. M. Anderson, T.-C. Tsao, and M. Levin, Adaptive Lift Control for a Camless Electrohydraulic Valvetrain, SAE Technical Paper 981029, 1998.
44. K. Misovec, B. Johnson, G. Mansouri, O. Sturman, and S. Massey, Digital Valve Technology Applied to the Control of a Hydraulic Valve Actuator, SAE Technical Paper 1999-01-0825, 1999.
45. W. Hoffmann and A.G. Stefanopoulou, Iterative Learning Control of Electromechanical Camless Valve Actuator, in *Proceedings of the American Control Conference*, Arlington, VA, 2001, pp. 2860–2866.
46. C. Tai and T.-C. Tsao, Control of an Electromechanical Camless Valve Actuator, in *Proceedings of the American Control Conference*, Anchorage, AK, 2002, pp. 262–267.
47. Z. Sun and T.-W. Kuo, Transient Control of Electro-Hydraulic Fully Flexible Engine Valve Actuation System, *IEEE Transactions on Control Systems Technology*, 18: 613–621, 2010.
48. M. Tomizuka, T.-C. Tsao, and K.-K. Chew, Analysis and Synthesis of Discrete-Time Repetitive Controllers, *Journal of Dynamic Systems, Measurement, and Control*, 111: 353–358, 1989.
49. T.-C. Tsao and M. Tomizuka, Robust Adaptive and Repetitive Digital Tracking Control and Application to a Hydraulic Servo for Noncircular Machining, *Journal of Dynamic Systems, Measurement, and Control*, 116: 24–32, 1994.
50. D.H. Kim and T.-C. Tsao, A Linearized Electrohydraulic Servovalve Model for Valve Dynamics Sensitivity Analysis and Control System Design, *Journal of Dynamic Systems, Measurement, and Control*, 122: 179–187, 2000.
51. P. Gillella, X. Song, and Z. Sun, Time-Varying Internal Model-Based Control of a Camless Engine Valve Actuation System, *IEEE Transactions on Control Systems Technology*, 22(4): 1498–1510, 2014.
52. F.Q. Zhao, D.L. Harrington, and M.C. Lai, *Automotive Direct-Injection Gasoline Engines*, Society of Automotive Engineers Press, Warrendale, PA, 2002.
53. F.J. Hamady, J.P. Hahn, K.H. Hellman, and C.L. Gray Jr., High-Speed/High-Resolution Imaging of Fuel Sprays from Various Injector Nozzles for Direct Injection Engines, SAE Technical Paper 950289, 1994.
54. K. Kawajiri, T. Yonezawa, H. Ohuchi, M. Sumida, and H. Katashiba, Study of Interaction between Spray and Air Motion, and Spray Wall Impingement, SAE Technical Paper 2002-01-0836, 2002.
55. M. Jansons, S. Lin, and K.T. Rhee, High-Speed Imaging from Consecutive Cycles, SAE Technical Paper 2001-01-3486, 2001.

56. J.D. Smith and V. Sick, Crank-Angle Resolved Imaging of Fuel Distribution, Ignition and Combustion in a Direct-Injection Spark-Ignition Engine, SAE Technical Paper 2005-01-3753, 2005.

57. P. Wyszynski, R. Aboagye, R. Stone, and G. Kalghatgi, Combustion Imaging and Analysis in a Gasoline Direct Injection Engine, SAE Technical Paper 2004-01-0045, 2004.

58. D.L.S. Hung, W. Humphrey, L. Markle, D. Chmiel, C. Ospina, and F. Brado, A Novel Transient Drop Sizing Technique for Investigating the Role of Gasoline Injector Sprays in Fuel Mixture Preparation, SAE Technical Paper 2004-01-1349, 2004.

59. D.L.S. Hung, G. Zhu, J. Winkelman, T. Stuecken, H. Schock, and A. Fedewa, A High Speed Flow Visualization Study of Fuel Spray Pattern Effect on Mixture Formation in a Low Pressure Direct Injection Gasoline Engine, SAE Technical Paper 2007-01-1141, 2007.

60. D.L.S. Hung, D.L. Harrington, A.H. Gandhi, L.E. Markle, S.E. Parrish, J.S. Shakal, H. Sayar, S.D. Cummings, and J.L. Kramer, Gasoline Fuel Spray Measurement and Characterization—A New SAE J2715 Recommended Practice, SAE Technical Paper 2008-01-1068, 2008.

61. J.D. Powell, M. Hubbard, and R.R. Clappier, Ignition Timing Controls Method and Apparatus, U.S. Patent 4063538, 1977.

62. G.M. Rassweiler and L. Withrow, Motion Picture of Engine Flames Correlated with Pressure Cards, *SAE Transactions*, 42: 185–204, 1938.

63. M.C. Sellnau, F.A. Matekunas, P.A. Battiston, C.-F. Chang, and D.R. Lancaster, Cylinder-Pressure-Based Engine Control Using Pressure-Ratio-Management and Low-Cost Non-Intrusive Cylinder Pressure Sensor, SAE Technical Paper 2000-01-0932, 2000.

64. R.J. Hosey and J.D. Powell, Closed Loop, Knock Adaptive Spark Timing Control Based on Cylinder Pressure, *Transactions of ASME*, 101, 1979.

65. K. Sawamoto, Y. Kawamura, T. Kita, and K. Matsushita, Individual Cylinder Knock Control by Detecting Cylinder Pressure, SAE Technical Paper 871911, 1987.

66. Y. Kawamura, M. Shinshi, H. Sato, N. Takahshi, and M. Iriyama, MBT Control through Individual Cylinder Pressure Detection, SAE Technical Paper 881779, 1988.

67. R. Muller, M. Hart, G. Krotz, M. Eickhoff, A. Truscott, A. Nobel, C. Cavalloni, and M. Gnielka, Combustion Pressure Based Engine Management System, SAE Technical Paper 2000-01-0928, 2000.

68. G.G. Zhu, C.F. Daniels, and J. Winkelman, MBT Timing Detection and Its Closed-Loop Control Using In-Cylinder Pressure Signal, SAE Technical Paper 2003-01-3266, 2003.

69. L. Eriksson, Spark Advance Modeling and Control, PhD Dissertation, Linkoping University, 1999.

70. L. Eriksson and L. Nielsen, Closed Loop Ignition Control by Ionization Current Interpretation, SAE Technical Paper 970854, 1997.

71. G. Zhu, I. Haskara, and J. Winkelman, Closed Loop Ignition Timing Control Using Ionization Current Feedback, IEEE Transactions on Control System Technology, 15(3), 2007.

72. G.G. Zhu, C.F. Daniels, and J. Winkelman, MBT Timing Detection and Its Closed-Loop Control Using In-Cylinder Ionization Signal, SAE Technical Paper 2004-01-2976, 2004.

73. C.F. Daniels, The Compression of Mass Fraction Burned Obtained from Cylinder Pressure Signal and Spark Plug Ion Signal, SAE Technical Paper 980140, 1998.

74. R.J. Hosey and J.D. Powell, Closed Loop, Knock Adaptive Spark Timing Control Based on Cylinder Pressure, *Transactions of ASME*, 101, 1979.

75. C.F. Daniels, G.G. Zhu, and J. Winkelman, Inaudible Knock and Partial Burn Detection Using In-Cylinder Ionization Signal, SAE Technical Paper 2003-01-3149, 2003.

76. G.G. Zhu, I. Haskara, and J. Winkelman, Stochastic Limit Control and Its Application to Knock Limit Control Using Ionization Feedback, SAE Technical Paper 2005-01-0018, 2005.

77. I. Haskara, G. Zhu, and J. Winkelman, IC Engine Retard Ignition Timing Limit Detection and Control Using In-Cylinder Ionization Signal, SAE Technical Paper 2004-01-2977, 2004.

78. N.A. Henein and M.K. Tagomari, Cold-Start Hydrocarbon Emissions in Port-Injected Gasoline Engines, *Progress in Energy and Combustion Science*, 25: 563–593, 1999.

79. J. Zhu and S.C. Chan, An Approach for Rapid Automotive Catalyst Light Off by High Values of Ignition Retard, *Journal of the Institute of Energy*, 167–173, 1996.

80. P. Tunestal, M. Wilcutts, A.T. Lee, and J.K. Hedrick, In-Cylinder Measurement for Engine Cold-Start Control, in *Proceedings of the IEEE International Conference on Control Applications*, Kohala Coast, HI, 1999, pp. 460–464.
81. J.D. Naber, J.R. Blough, D. Frankowski, M. Goble, and J.E. Szpytman, Analysis of Combustion Knock Metrics in Spark-Ignited Engines, SAE Technical Paper 2006-01-0400, 2006.
82. I. Haskara, G. Zhu, and J. Winkelman, Multivariable EGR/Spark Timing Control for IC Engines via Extremum Seeking, presented at Proceedings of the American Control Conference, Minneapolis, MN, 2006.

3

Design, Modeling, and Control of Automotive Transmission Systems

3.1 Introduction to Various Transmission Systems

The basic function of a transmission is to transit power from the engine to the wheels of a vehicle. The fundamental reason why we need a transmission between the engine and the vehicle is again attributed back to the dynamic nature of vehicle operation. First, the vehicle requires both forward and reverse motion, while the engine only works in one direction. Therefore, a transmission that can reserve the output direction is needed. Second, the engine power, efficiency, and emissions are functions of the engine speed and load. To deliver the maximum power to the vehicle and maintain high fuel efficiency, different speed ratios are required between the engine and the vehicle. Third, it is desirable to reduce the driveline vibration caused by the engine firing pulse or the road disturbances. This is often referred to as drivability, which is a key attribute of the transmission system. Given the above reasons, the transmission is an indispensable part of the automotive propulsion system.

To realize the above functions, various transmissions have been designed, including the manual transmission (MT), automated manual transmission (AMT), dual-clutch transmission (DCT), automatic transmission (AT), continuously variable transmission (CVT), etc. Both MT and AMT employ countershaft gears to realize different gear ratios. The DCT uses two sets of countershaft gears where the odd gears are on one set and the even gears on the other set. The countershaft gears are fairly flexible for designing the desired gear ratios and are extremely efficient for transmitting mechanical power, but they are not very compact given the fact that there are two parallel shafts and the gears are located along the axial direction of the shafts. The AT uses planetary gear sets to change the speed ratio. This is accomplished by connecting different nodes of the planetary gear sets (for details see Section 3.2). The planetary gears are fairly compact since they are concentric. But the gear ratio design is much more complicated than the MT. The CVT has several different variations: the belt system, the chain system, and the toroidal drive system. The belt and chain drive systems are suitable for low to medium torque applications, while the toroidal drive system has higher torque capacity. But overall the design of the CVT is more complicated than the stepped gear transmission.

Besides the different gear ratio mechanisms employed by different transmissions, the other key differentiator for the transmissions is the actuation and control methods for realizing the gear ratios in real time. The MT has the driver to control the gear ratio change by using a lever mechanism. A synchronization device is used to synchronize the speeds of the input shaft and the output shaft before a locking device is used to lock the shafts into the new gear ratio. The AT uses electrohydraulic clutches to realize the gear shift automatically. The gear shift schedule determines the timing for gear shift based on the vehicle speed and the throttle position. A hydraulic pump is connected directly to the engine that will produce the fluid power for actuating the clutches. Control valves are used to regulate the pressure and flow for appropriate gear shift. Frictional material is used for the clutch plate, and when compressed will engage the input and output shafts. The CVT also employs fluid power to change the speed ratio automatically in real time. But the control mechanism is more complicated due to the nature of continuous gear ratio change. The AMT uses either an electrical motor or a hydraulic actuator to change the gear ratio. The DCT has two clutches that can be engaged or disengaged in real time to realize the gear ratio change. One clutch is connected to the odd gears, and the other clutch is connected to the even gears. During gear shift, synchronization between the clutches is critical. For the rest of the chapter, we will mainly focus on the automatic transmission, while many results can also be applied to other types of transmissions.

3.2 Gear Ratio Realization for Automatic Transmission

3.2.1 Planetary Gear Set

Gear ratio change in an automatic transmission is realized by connecting different nodes of the planetary gear sets. As shown in Figure 3.1, a planetary gear set includes the sun gear, the ring gear, and the planet gear. The carrier connects the planet gears together. So the three nodes for a planetary gear set are sun, ring, and carrier.

To analyze the speed and torque of a planetary gear set, we use the lever diagram as shown in Figure 3.2, where SG represents the sun gear, C represents the carrier, and RG represents the ring gear. The length of the lever between SG and C is denoted as R, which

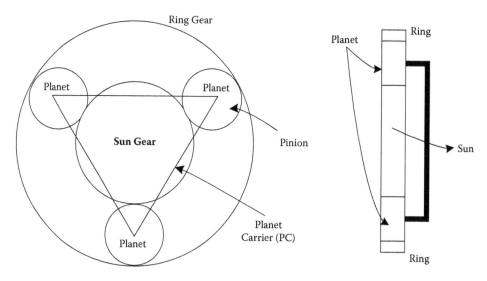

FIGURE 3.1
Diagram of a planetary gear set.

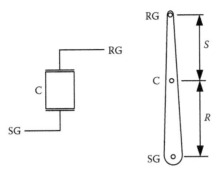

FIGURE 3.2
Lever diagram.

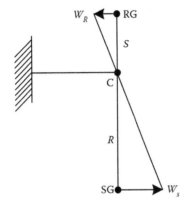

FIGURE 3.3
Speed calculation using lever diagram.

corresponds to the number of teeth of the ring gear. Similarly, the length between RG and C is S, which corresponds to the number of teeth of the sun gear.

Let's connect the carrier (node C) to the ground as shown in Figure 3.3. Now we are ready to apply the lever analogy to calculate the speed and torque at different nodes of the planetary gear set.

Based on Figure 3.3, we have

$$\frac{W_R}{W_S} = -\frac{S}{R}$$

Based on Figure 3.4, we have

$$\begin{cases} T_R \cdot S = T_S \cdot R \\ T_R + T_S - T_C = 0 \end{cases} \Rightarrow T_R = T_S \frac{R}{S}, \quad T_C = T_S + T_R = \frac{S+R}{S} T_S$$

With the above example, we introduce the following relationship for the planetary gear set:

$$T_s + T_R - T_C = 0$$

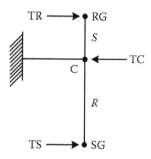

FIGURE 3.4
Torque calculation using lever diagram.

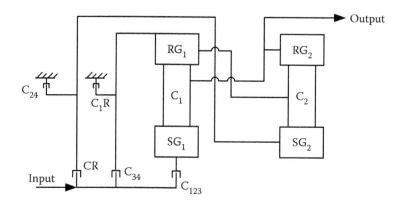

FIGURE 3.5
Diagram of a four-speed automatic transmission.

$$T_R S = T_S R \tag{3.1}$$

$$W_s S + W_R R = W_C (R + S)$$

So there are two degrees of freedom for the speeds and one degree of freedom for the torques. In other words, only after two speeds are determined, can the third speed be calculated. But if one torque is determined, the remaining torques can be calculated.

3.2.2 Speed and Torque Calculation for Automatic Transmission

In automatic transmissions, to obtain enough gear ratios, two or more planetary gear sets are employed. The diagram for a four-speed automatic transmission is shown in Figure 3.5. The lever diagram and shift table for the transmission are shown in Figure 3.6 and Table 3.1, respectively.

Before calculating the speed and torque ratios for the transmission, we show how the lever diagram for the compounded planetary gear sets is obtained. Figure 3.7 shows the compound gear used in the automatic transmission. Figure 3.8 shows the corresponding lever diagram.

Combine the two levers in Figure 3.8 into one, and scaling the length between S_{G2} and C_2 accordingly, we have the new lever diagram as shown in Figure 3.9.

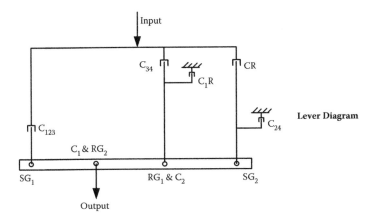

FIGURE 3.6
Lever diagram of a four-speed automatic transmission.

TABLE 3.1

Shift Table for the Automatic Transmission

Gear	C_{123}	C_1R	C_{24}	C_{34}	CR
1	X	X			
2	X		X		
3	X			X	
4			X	X	
R		X			X

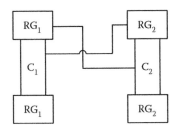

FIGURE 3.7
Compound planetary gear set.

Recall the speed relationship for the planetary gear sets:

$$W_{S1}S_1 + W_{R1}R_1 = W_{C1}(R_1 + S_1)$$

$$W_{S2}S_2 + W_{R2}R_2 = W_{C2}(R_2 + S_2)$$

$$W_{R1} = W_{C2}$$ (3.2)

$$W_{R2} = W_{C1}$$

For first gear (see Table 3.1), S_{G1} is the input, C_1 is the output, and C_2 is grounded. We have $W_{R1} = W_{C2} = 0$, $W_{S1} = W_{in}$, $W_{R2} = W_{C1} = W_{out}$.

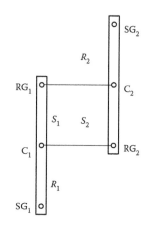

FIGURE 3.8
Lever diagram of the compound gear.

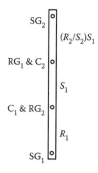

FIGURE 3.9
Combined lever diagram for the compound gear.

Plugging them into the above equations, we have

$$W_{in}S_1 + 0 = W_{out}(R_1 + S_1)$$

$$W_{S2}S_2 + W_{OUT}R_2 = 0$$

$$\therefore W_{out} = \frac{S_1}{R_1 + S_1}W_{in}$$

$$W_{S2} = -\frac{R_2}{S_2}W_{out} = -\frac{R_2 S_1}{S_2(R_1 + S_1)}W_{in} \qquad (3.3)$$

Let's check the geometric relationship between the speeds from the lever diagram (Figure 3.10):

$$W_{out} = W_{in}\frac{S_1}{R_1 + S_1}$$

$$W_{S2} = -\frac{R_2 S_1}{S_2(R_1 + S_1)}W_{in}$$

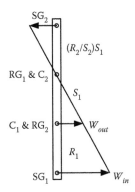

FIGURE 3.10
Speed calculation for the compound gear.

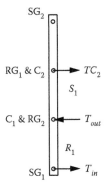

FIGURE 3.11
Torque calculation for the compound gear.

Now let's calculate the torque relationship at first gear (Figure 3.11):

$$T_{in}(R_1 + S_1) = T_{out} S_1$$

$$T_{C2} = T_{in} - T_{out}$$

$$\therefore T_{out} = T_{in} \frac{R_1 + S_1}{S_1}$$

$$T_{C2} = T_{in} - \frac{R_1 + S_1}{S_1} T_{in} = \frac{R_1}{S_1} T_{in}$$

(3.4)

The lever diagram for the second gear is shown in Figure 3.12. Using similar methods as for the first gear, we can calculate the speed ratio for the second gear:

$$\frac{W_{in}}{W_{out}} = \frac{R_1 + S_1 + R_2 S_1 / S_2}{S_1 + R_2 S_1 / S_2}$$

$$\frac{W_{in}}{W_{out}} = \frac{R_1 S_2 + S_1 S_2 + R_2 S_1}{S_1 S_2 + R_2 S_1}$$

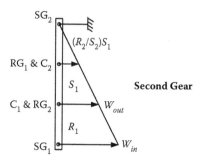

FIGURE 3.12
Lever diagram for second gear.

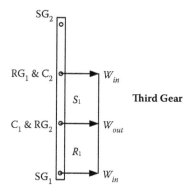

FIGURE 3.13
Lever diagram for third gear.

Similarly, the lever diagram for the third gear is shown in Figure 3.13.

$$W_{out} = W_{in}$$

The lever diagram for the fourth gear is shown in Figure 3.14.

$$\frac{W_{in}}{W_{out}} = \frac{\dfrac{R_2 S_1}{S_2}}{R_2 \dfrac{S_1}{S_2} + S_1} = \frac{R_2 S_1}{R_2 S_1 + S_1 S_2} = \frac{R_2}{R_2 + S_2}$$

The lever diagram for the reverse gear is shown in Figure 3.15.

$$\frac{W_{in}}{W_{out}} = -\frac{\dfrac{R_2 S_1}{S_2}}{S_1} = -\frac{R_2}{S_2}$$

The torque ratio can be calculated similarly. Also, given the conservation of energy, the torque ratio will be the inverse of the speed ratio. Recall $\dfrac{T_{in}}{T_{out}} = \dfrac{S_1}{R_1 + S_1}$ for the first gear.

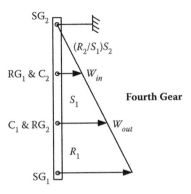

FIGURE 3.14
Lever diagram for fourth gear.

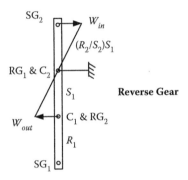

FIGURE 3.15
Lever diagram for reverse gear.

3.2.3 Speed and Torque Calculation during Gear Shifting

In this section, we are going to discuss the clutch shift control during the gear ratio change. Recall the stick diagram and the clutch map of the four-speed automatic transmission (see Figure 3.5).

Recall the basic equations for a planetary gear set (see Equation (3.1)):

$$T_s + T_R = T_C$$

$$T_R S = T_S R$$

$$W_s S + W_R R = W_C (R + S)$$

For the four-speed automatic transmission, at first gear, C_{123} and C_{1R} are engaged. The torque relationship is shown in Figure 3.16.

$$T_{in}(R_1 + S_1) = T_{out} S_1$$

$$T_{C1R} = -T_{in} + T_{out}$$

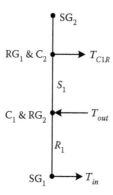

FIGURE 3.16
Torque for first gear.

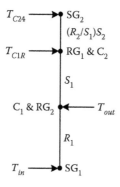

FIGURE 3.17
Torque during first to second gear shift.

$$\therefore T_{out} = T_{in} \frac{R_1 + S_1}{S_1}$$

$$T_{C1R} = \frac{R_1}{S_1} T_{in}$$

Now if we shift from first gear to second gear, clutch C_{24} is engaged and clutch C_{1R} is disengaged. There is a time period where both clutch C_{24} and clutch C_{1R} are interacting with the planetary gear set as shown in Figure 3.17.

$$T_{C24} + T_{C1R} + T_{in} = T_{out}$$

$$T_{C24} \frac{R_2 S_1}{S_2} + T_{out} S_1 = T_{in}(R_1 + S_1)$$

$$\therefore T_{out} = T_{in} \frac{R_1 + S_1}{S_1} - T_{C24} \frac{R_2}{S_2}$$

$$T_{C1R} = T_{out} - T_{in} - T_{C24} = T_{in} \frac{R_1}{S_1} - T_{C24} \left(1 + \frac{R_2}{S_2}\right)$$

Eventually, when clutch C_{1R} is fully disengaged, i.e., $T_{C1R} = 0$, we have

$$T_{C24} = T_{in} \frac{R_1 S_2}{S_1(R_2 + S_2)}$$

$$T_{out} = T_{in} \frac{R_1 S_2 + R_2 S_1 + S_1 S_2}{S_1 S_2 + S_1 R_2}.$$

(3.5)

This time period is called the torque phase of the clutch shift.

The next phase during the clutch shift is called the inertia phase, where the input/engine speed needs to be brought back to the speed set by the new gear ratio, as shown in Figure 3.18. So there will be a speed change. Let the engine speed deceleration be α; then the input torque becomes $T_{in} + I_{in}\alpha$. Based on the level diagram in Figure 3.19, we have

$$T_{C24} = (T_{in} + I_{in}\alpha) \frac{R_1 S_2}{S_1(R_2 + S_2)}$$

$$T_{out} = (T_{in} + I_{in}\alpha) \frac{R_1 S_2 + R_2 S_1 + S_1 S_2}{S_1 S_2 + S_1 R_2}$$

(3.6)

The complete diagram for the gear shift from first to second gear is shown in Figure 3.20, where the input speed, output torque, and clutch torques are illustrated.

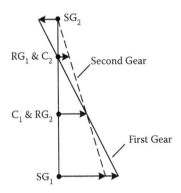

FIGURE 3.18
Speed changes from first gear to second gear.

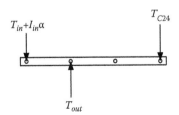

FIGURE 3.19
Lever diagram during inertia phase.

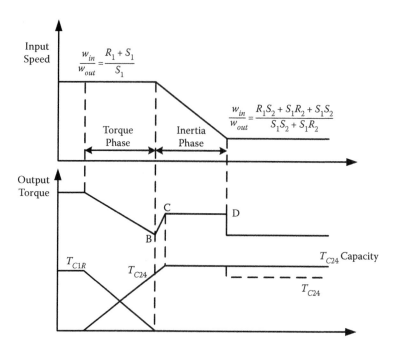

FIGURE 3.20
Clutch-to-clutch shift during first to second gear.

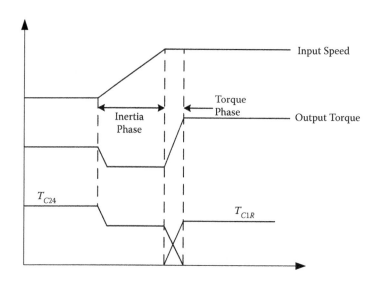

FIGURE 3.21
Clutch-to-clutch shift during second to first gear.

During the gear ratio change, the two clutches need to be synchronized. If T_{C1R} is released too early, the engine will flare. If T_{C1R} is released too late, there will be a tire-up. The output torque will drop.

Now let's look at the downshift (2→1). This process is exactly the opposite of the upshift. As shown in Figure 3.21, the inertia phase will occur first by reducing the torque on clutch C_{24}, and the engine speed will rise to the speed set by the new gear ratio. Then the clutch torque for C_{24} will be reduced further and the torque for C_{1R} will be increased in a synchronized fashion. This will complete the torque phase.

3.3 Design and Control of Transmission Clutches

From Section 3.2, it is clear that the gear ratios for automatic transmissions are realized by connecting different nodes of the planetary gear sets. This is achieved using electro-hydraulically actuated clutches. More importantly, for the clutch-to-clutch shift technology, where the oncoming clutch needs to be engaged and the offgoing clutch needs to be disengaged, synchronization between the two clutches is critical (see Figures 3.20 and 3.21). Therefore, clutch is a critical element of the transmission system, and its performance directly affects the overall functionality and performance of the transmission. This section will present the design, modeling, and control of transmission clutches.

3.3.1 Clutch Design

The commonly used clutch actuation devices in automatic transmissions are electro-hydraulically actuated clutches [1]. This is mainly due to the high power density of the fluid power system. A schematic diagram of a typical transmission clutch actuation system is shown in Figure 3.22 [1], while an illustration of the physical system is shown in Figure 3.23 [3]. When the clutch needs to be engaged, pressurized fluid flows into the clutch chamber and pushes the clutch piston toward the clutch packs until they are in contact. This process is called clutch fill. There is no torque transmission yet through the clutch at this stage. When the clutch is ready to transmit torque, the input pressure is further increased, which then squeezes the clutch packs with the clutch piston. The clutch packs are circular disks, and they are connected to different input and output shafts. Engagement of the clutch packs will enable the transfer of the engine torque to the vehicle driveline.

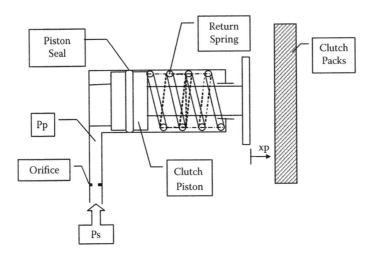

FIGURE 3.22
Schematic diagram of a clutch mechanism.

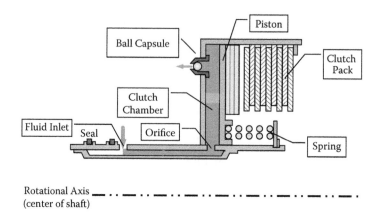

FIGURE 3.23
Physical illustration of a clutch mechanism.

For the clutch-to-clutch shift technology described in Section 3.2, it is imperative to control the clutch piston to reach the clutch packs within a specified clutch fill time because an improper clutch fill process can result in either an *underfill* or an *overfill* [2], both of which can cause the failure of the clutch shift synchronization, and therefore negatively affect the clutch shift quality. Underfill refers to the case where the clutch piston has not reached the clutch packs at the end of the clutch fill process. So the clutch is not ready for engagement when it is supposed to be. Overfill refers to the case where the clutch piston has traveled more than required and started to squeeze the clutch packs at the end of the clutch fill process. This will start torque transmission prematurely. The clutch engagement process will see a significant pressure rise by squeezing on the clutch packs to transmit torque. The smooth and synchronized torque transmission is critical for shift and drive quality. In the following sections, several different methods will be introduced to control the clutch fill and engagement processes.

3.3.2 New Clutch Actuation Mechanism

In this section, we show a new clutch design that will automatically achieve precise clutch fill control [3]. Figure 3.24(a) [3] shows the simplified schematic diagram of the clutch actuation system, which mainly consists of two on/off valves, an internal feedback spool (IFS), an IFS return spring, a feedback channel, and the clutch assembly. The IFS controls the opening orifice between the supply pressure and the clutch chamber, and is the key component of the clutch actuation mechanism.

The following is the operating procedure of the system:

1. When the clutch piston is at the initial disengaged position (Figure 3.24(a)), on/off valve 1 connects the IFS orifice, A_{IFS}, to the fluid tank. The pressure inside the clutch chamber is low, so the clutch piston will be kept at the disengaged position by the piston return spring.

2. When the clutch fill process begins, on/off valve 1 is energized to connect the IFS orifice to the supply pressure. Pressurized fluid then enters the clutch chamber through the IFS orifice, overcomes the spring force, and pushes the clutch piston to the right. On the other side, on/off valve 2 at the feedback exhaust port is closed

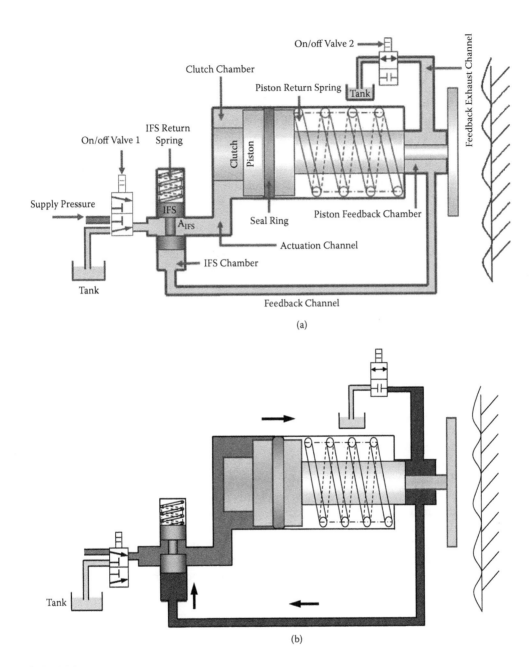

FIGURE 3.24
(a) Schematic diagram of the clutch actuation system. (b) Piston and IFS motion during the clutch fill. (From X. Song et al., *IEEE/ASME Transactions on Mechatronics*, 17(3): 582–587, 2012. With permission.)

to block the fluid inside the piston feedback chamber from flowing out to the tank. Therefore, the motion of the clutch piston will squeeze the fluid in the piston feedback chamber to the IFS chamber through the feedback channel, which then drives the IFS upward (Figure 3.24(b) [3]). As the IFS moves upward, the IFS orifice area gradually decreases. This restricts the flow through the IFS orifice, which then slows down the pressure rise in the clutch chamber. The regulated chamber pressure will again affect the clutch piston displacement, and therefore form a feedback loop. The clutch piston and the IFS continue to move in a synchronized fashion until the IFS orifice is cut off, which then separates the clutch chamber from the supply pressure, and the piston naturally stops at the predetermined position.

From the above analysis, it can be observed that the clutch piston displacement is fed back through the IFS orifice area, which regulates the fluid pressure inside the clutch chamber. The flow, pressure, and displacement are precisely coordinated by the embedded hydromechanical feedback loop. This internal feedback structure ensures robust and precise motion of the clutch fill process. The feedback structure also guarantees a smoothly decreasing clutch piston velocity profile as it approaches the clutch fill final position.

3. After the clutch fill, the clutch packs are ready to be engaged for torque transmission. When the clutch engagement starts, the clutch piston needs to be pushed further to squeeze the clutch packs. This can be realized by opening on/off valve 2 at the exhaust port to release the fluid in the feedback chamber to the tank. As a result, the pressures in the piston feedback chamber and the IFS chamber will drop, and subsequently the IFS return spring pushes the IFS downward until the IFS orifice is fully open. The supply pressure, P_s, which now gets reconnected to the clutch chamber, will push the clutch piston further to squeeze the clutch packs.

4. During the clutch disengagement, the clutch piston needs to move back to its initial position so that the clutch packs will be released. On/off valve 1 is deenergized to connect the IFS orifice to the low-pressure tank. The piston return spring will then push the piston back to the disengaged position.

The control block diagram of the IFS clutch actuation system is shown in Figure 3.25 [3], which shows that the clutch mechanism consisting of the IFS, the clutch piston, and the on/off valve can be represented in a feedback loop.

There are two advantages of this new mechanism for clutch control. First, instead of designing a controller considering the complex nonlinear hydraulic dynamics, the proposed IFS clutch actuation system can realize a fast, precise, and robust nonlinear control by self-regulating the IFS spool orifice with the hydromechanical feedback rather than sensor measurement. The orifice area automatically regulates the clutch chamber pressure, and thus enables a smooth and precise clutch piston velocity and displacement profile.

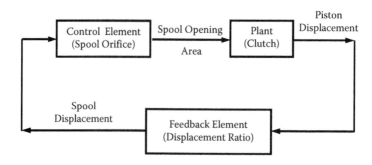

FIGURE 3.25
Feedback control diagram for the clutch mechanism. (From X. Song et al., *IEEE/ASME Transactions on Mechatronics*, 17(3): 582–587, 2012. With permission.)

Second, ensuring a precise and robust clutch fill with the IFS system at the clutch level will enable the simplification of the upstream hydraulic control system, including the control valves, the electronic control devices, and their accessories [4]. This not only will improve the transmission system compactness, but also can alleviate the external control calibration efforts.

3.3.2.1 Simulation and Experimental Results

The new clutch actuation mechanism is fabricated for experimental verification. The test bed is shown in Figure 3.26 [3]. The main components include a servo motor, an automotive transmission pump, a pilot-operated proportional relief valve, two on/off valves, a flow meter, two pressure sensors, a clutch mounting device/fixture, a displacement sensor, a power supply unit with servo amplifier, and an xPC Target real-time control system. In particular, one on/off valve controls the supply oil into the clutch chamber, and the other controls the hydraulic oil injection into the internal feedback chamber for back pressure. The sensors in the test bed are not used for the clutch control, but only for performance verification purposes since the IFS system operation requires no measurement feedback.

The clutch fill experimental results are shown in Figures 3.27 to 3.30 [3]. Note that the pressures shown are absolute pressure. The supply pressure is kept at 12 bar using a pressure relief valve. On/off valve 1 is initially closed to separate the clutch chamber from the high-pressure port, while on/off valve 2 is open. After injecting hydraulic oil into the internal feedback channel for back pressure, the internal feedback IFS chamber is cut off by closing on/off valve 2. Then on/off valve 1 is opened to connect the supply pressure with the input chamber, and thus high-pressure fluid can flow in to the clutch chamber and push the clutch piston forward. Note that the operations of the two on/off valves are completely separate in time, and the transients of the two valves will not affect each other in the clutch fill. More than 30 groups of clutch fill tests have been conducted, and all of them have shown good repeatability. The clutch piston displacements from six representative tests are shown in Figure 3.27, which clearly shows that the displacement trajectories from different

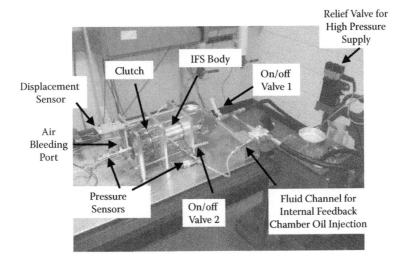

FIGURE 3.26
IFS clutch actuation test bed. (From X. Song et al., *IEEE/ASME Transactions on Mechatronics*, 17(3): 582–587, 2012. With permission.)

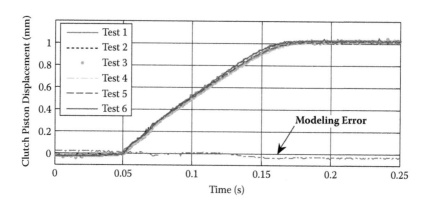

FIGURE 3.27
Multiple tests for clutch piston displacement and the dynamics modeling error of the clutch piston motion. (From X. Song et al., *IEEE/ASME Transactions on Mechatronics*, 17(3): 582–587, 2012. With permission.)

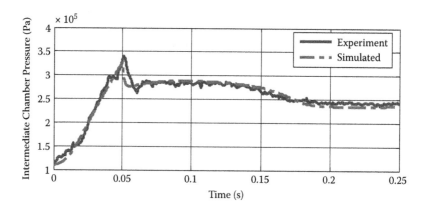

FIGURE 3.28
Intermediate actuation chamber pressure. (From X. Song et al., *IEEE/ASME Transactions on Mechatronics*, 17(3): 582–587, 2012. With permission.)

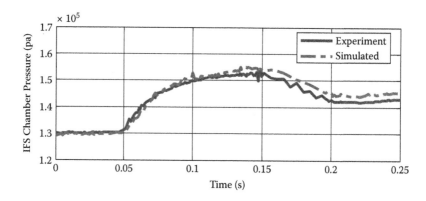

FIGURE 3.29
IFS chamber pressure. (From X. Song et al., *IEEE/ASME Transactions on Mechatronics*, 17(3): 582–587, 2012. With permission.)

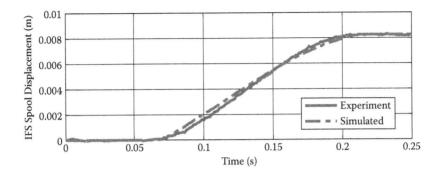

FIGURE 3.30
IFS spool displacement. (From X. Song et al., *IEEE/ASME Transactions on Mechatronics*, 17(3): 582–587, 2012. With permission.)

tests can almost overlap each other. Compared with the open-loop clutch control [2] in the current production vehicle, the new clutch mechanism is more robust. The fundamental reason is the feedback arrangement enabled by the internal feedback system (IFS).

At the start of the clutch fill process, the pressure in the intermediate actuation chamber, which is connected to the clutch chamber through a channel, quickly goes up due to the incoming high-pressure fluid, as shown in Figure 3.28. At 0.05 s, the pressure in the intermediate actuation chamber (Figure 3.28), as well as in the clutch chamber, is high enough to overcome the piston return spring force to push the clutch piston forward, and subsequently the clutch chamber pressure drops due to the piston motion-induced outflow. The pressure inside the clutch chamber is then kept at an appropriate level to keep moving the clutch piston forward. The pressure in the piston feedback and IFS chamber will then go up (Figure 3.29) due to the clutch piston motion, and thus pushes the IFS spool to close the IFS orifice. Once the IFS spool travels to the end and therefore cuts off the IFS orifice, as shown in Figure 3.30, the clutch piston stops at around 0.175 s. At the end of the clutch fill, the clutch piston travels 1 mm robustly, as shown in Figure 3.28, which is exactly the desired clutch fill final displacement. The whole clutch fill process finishes within 175 ms, which is again evidenced by the chamber pressure transient in Figure 3.28.

3.3.3 Feedforward Control for Clutch Fill

In this section, we present the modeling of the transmission clutch (Figure 3.22 [1]) and a feedforward control for the clutch fill process. There are two main challenges associated with the clutch fill control. First, even small errors in calculating the clutch fill pressure and fill time could lead to an overfill or an underfill, which will adversely impact the shift quality. Second, currently there is no pressure sensor inside the clutch chamber, and therefore a pressure feedback control loop cannot be formed. So, it is necessary to design an open-loop pressure control profile, which should be optimal in the sense of peak flow demand and also robust in terms of clutch fill time. Clearly, the traditional approach based on manual calibration is not effective to achieve the above objectives. In this section, we will present a systematic approach to solve this problem [1].

To enable a systematic and model-based control design, a precise clutch fill model is necessary. It should capture the key dynamics of the clutch fill process, which are not identical to those of the clutch engagement process [5–7]. Based on the clutch fill dynamic model, a systematic approach for this control problem can be determined recursively via Bellman's dynamic programming (DP) method [8, 9]. However, different from other applications, the clutch actuation system contains high frequency and stiff dynamics [6, 7], which requires a very fast sampling rate for the discretized model. As a result, the large number of steps, along with the curse of dimensionality [10] associated with conventional numerical DP method, prohibits its efficient implementation for the clutch fill control problem. Furthermore, the conventional numeric DP algorithm suffers from interpolation errors, which will affect the accuracy of the final results. Therefore, we will show a customized numerical dynamic programming approach, which has successfully solved the above problems and generated satisfactory results. The customized DP method transforms the stiff dynamic model into a nonstiff one by a unique model structure transformation and state discretization, and thus enables the reduction of the discrete sampling steps. Furthermore, by discretizing the state space into regions rather than discrete nodes, the interpolation error is avoided. Finally, the customized DP only searches the optimal solution within the reachable discrete states, which dramatically reduces the searching space and therefore mitigates the curse of the dimensionality problem [10].

3.3.3.1 Clutch System Modeling

As shown in Figure 3.22, p_s is the supply pressure command and also the control input to the system, p_p is the pressure inside the clutch chamber, and x_p is the clutch piston displacement. The pressurized fluid flows into the clutch chamber and pushes the clutch piston to the right, and finally contacts the clutch packs. The dynamics associated with clutch fill can be modeled as [1]

$$\dot{x}_1 = x_2 \tag{3.7}$$

$$\dot{x}_2 = \frac{1}{M_p}[A_p \times (x_3 + P_c - P_{atm}) - D_p x_2 - F_{drag}(x_3 + P_c, x_2) - K_p \times (x_1 + x_{p0})] \tag{3.8}$$

$$\dot{x}_3 = \frac{\beta}{V_0 + A_p x_1}\left[sign(u - x_3)C_d A_{orifice}\sqrt{\frac{2|u - x_3|}{\rho}} - A_p x_2 \right] \tag{3.9}$$

where

$$\begin{bmatrix} x_1 = x_p \\ x_2 = \dot{x}_p \\ x_3 = p_p \\ u = p_s \end{bmatrix}$$

and

$$\begin{bmatrix} x_1(0) \\ x_2(0) \\ x_3(0) \end{bmatrix} = \begin{bmatrix} 0 \\ 0 \\ u(0) \end{bmatrix}$$

where x_1 is the clutch piston displacement, x_2 is the clutch piston velocity, x_3 is the clutch chamber pressure, u is the supply pressure control input, M_p is the effective mass of the piston, A_p is the piston surface area, D_p is the clutch damping coefficient, P_{atm} is the atmospheric pressure, K_p is the piston return spring constant, x_{p0} is the return spring preload, β is the fluid bulk modulus, V_0 is the chamber volume, C_d is the discharge coefficient, $A_{orifice}$ is the orifice area, and ρ is the fluid density. F_{drag} is the piston seal drag force, which is dependent on the piston motion and modeled as

$$F_{drag} = \begin{cases} \left[k_m (x_3 + P_c) + c_m \right] \times \tanh\left(\dfrac{x_2}{\alpha} \right) & (x_2 \neq 0) \\ F_{stick} & (x_2 = 0) \end{cases} \tag{3.10}$$

where k_m and c_m are constant, α is the piston seal damping coefficient, and F_{stick} is the static stick friction force from the Kanopp's stick-slip model [11]. The stick friction is often neglected in the clutch dynamic models for engagement [5–7], as it is relatively small compared with the clutch engagement force. But due to the low operating pressure during the clutch fill, the stick friction force becomes critical. For numerical stability, it is assumed that the drag force is F_{stick} when the piston velocity x_2 is within a small interval around zero [11], which is called the stick region, as shown in Figure 3.31 [1]. When the velocity is in this region, the value of F_{stick} is to balance the net force and the piston acceleration is assumed to be zero. Moreover, there is a maximum value constraint for the stick friction. If the net force exceeds the maximum stick friction, the piston will accelerate. The maximum stick friction, which is noted as F_{static}, is proportional to the clutch chamber pressure and can be modeled as

$$F_{static} = k_s(x_3 + P_c) + c_s \tag{3.11}$$

where k_s and c_s are constant.

In addition, P_c is the centrifugal force-induced pressure generated from the rotation of the clutch assembly [12]. The fluid pressure distribution P_{ct} due to the centrifugal force at any radius r can be expressed as

$$P_{ct} = \frac{\rho}{2} \omega^2 \left(r^2 - r_{st}^2 \right) \tag{3.12}$$

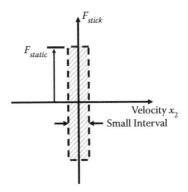

FIGURE 3.31
Stick friction diagram. (From X. Song et al., *Journal of Dynamic Systems, Measurement, and Control*, 133: 054503, 2011. With permission.)

where ω is the clutch system rotational speed and r_{st} is the starting fluid level [13]. The average fluid centrifugal pressure P_c on the effective piston area A_p can be expressed as

$$P_c = \frac{\displaystyle\int_{r_{pi}}^{r_{po}} P_{ct} \cdot 2\pi r dr}{A_p} \tag{3.13}$$

where r_{pi} and r_{po} are the piston inner and outer radiuses, respectively.

3.3.3.2 Formulation of the Clutch Fill Control Problem

To enable a fast and precise clutch fill, the clutch piston must travel exactly the distance d, which is required for the piston to contact the clutch packs, in the desired clutch fill time T. Also, at time T, the piston velocity x_2 should be zero, and the pressure force inside the chamber must be equal to the spring force in order to keep the piston in contact with the clutch packs. These requirements can be translated into a set of final conditions that the system must satisfy [1]:

$$x_1(T) = d, \quad x_2(T) = 0 \quad \text{and} \quad x_3(T) = \frac{K_p \times (d + x_{p0})}{A_p} + P_{atm} - P_c \tag{3.14}$$

where d is the desired clutch stroke, and $x_1(T)$, $x_2(T)$, and $x_3(T)$ are the final states.

Among the controls that can bring the clutch from initial states to the final states (3.14), we would like to take the one that has minimum peak flow demand. This will enable a smaller displacement transmission pump, which in turn will improve the vehicle fuel economy and reduce cost [4]. To reduce the peak flow demand, we need to minimize the peak value of the piston velocity x_2 during the clutch fill process since the clutch fill flow is proportional to the piston velocity. Therefore, x_2 should quickly approach the average velocity, and stay at the average velocity as long as possible, and at the end goes to the final state conditions as shown in Figure 3.32 [1]. Note that the total area enclosed by the x_2 trajectory should be the piston displacement d.

In addition, the piston velocity cannot increase too fast at the beginning of the clutch fill. Before the clutch starts moving, the clutch chamber pressure increases, and therefore the

Piston Velocity Versus Time

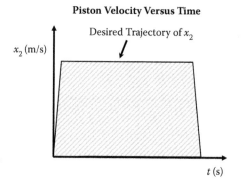

FIGURE 3.32
Desired trajectory of x_2. (From X. Song et al., *Journal of Dynamic Systems, Measurement, and Control*, 133: 054503, 2011. With permission.)

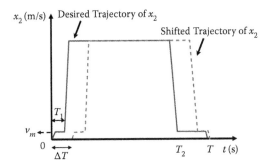

FIGURE 3.33
Shifted trajectory of x_2. (From X. Song et al., *Journal of Dynamic Systems, Measurement, and Control*, 133: 054503, 2011. With permission.)

stick friction on the clutch piston also increases. The stick friction will reach its maximum value F_{static}, and then the clutch piston will move. Once moving, the piston drag force will switch from stick friction to drag force in motion, which is smaller than the F_{static}. This friction force transient is nonsmooth and nonlinear [11], which makes it difficult to track and control a fast-changing piston velocity profile, especially for open-loop control. Therefore, in Figure 3.33 [1], the initial velocity of the clutch fill is designed to be small for a short duration and then rise quickly to its peak value.

However, the above considerations have not taken system robustness into account. In particular, the solenoid valve, which is used to generate the input pressure command u, has time delay and subsequently results in the shift of the x_2 trajectory, as shown in Figure 3.33. Note that the final time T is fixed, so the piston could not travel to the desired distance d due to the shift, and the difference between the desired trajectory and the shifted one will be

$$\Delta d = \int_{T-\Delta T}^{T} x_2(t)\,dt \tag{3.15}$$

Therefore, to minimize Δd, the value of $x_2(t)$ between time $T - \Delta T$ and T should be as small as possible, as shown in Figure 3.33, and we can claim that the unique trajectory of x_2 will enhance the robustness of the system for time delay. In addition, from (3.14), we can see that the clutch fill final states are determined by the spring stiffness K_p, the piston area A_p, the spring preload x_{po}, and the centrifugal pressure P_c. K_p, A_p, and x_{po} can be measured accurately and will not change much with the environment. P_c can also be accurately determined based on the transmission rotational speed, and it in fact has less influence on the final condition due to its small magnitude compared with other forces. Therefore, the final states of the clutch fill system are quite robust.

Now we are ready to formulate the clutch fill control as an optimization problem. The cost function of the optimization problem is

$$g = \int_0^{T_1} [x_2(t) - v_m]^2\,dt + \lambda_1 \int_{T_1}^{T_2} \left[x_2(t) - \frac{d}{T}\right]^2 dt + \lambda_2 \int_{T_2}^{T} [x_2(t) - v_m]^2\,dt + \lambda_3 [x_1(T) - d]^2$$

$$+ \lambda_4 [x_2(T) - 0]^2 + \lambda_5 \left\{ x_3(T) - \left[\frac{K_p(d + x_{p0})}{A_p} + P_{atm} - P_c\right] \right\}^2 \tag{3.16}$$

In particular, the first term of the cost function ensures the piston to start with a low velocity v_m. The second term ensures that the velocity x_2 remains close to the average velocity d/T, which will minimize the peak value of x_2, and therefore the peak flow demand. The third term ensures x_2 to be as small as possible (close to zero) from time T_2 to final time T, which will enhance system robustness. The last three terms ensures that the system will reach the specified final conditions in the required time T. λ_1, λ_2, λ_3, λ_4, and λ_5 are the weighting factors.

3.3.3.3 Optimal Control Design

A systematic solution to the above optimization problem can be determined recursively via Bellman's dynamic programming. Since the system model (Equations (3.7) to (3.9)) is nonlinear, an analytical solution cannot be obtained. Instead, a numerical solution will be provided. But first we need to discretize the system model to carry out the numerical dynamic programming method.

3.3.3.3.1 System Model Discretization

We discretize the system model (Equations (3.7) to (3.9)) as follows [1]:

$$x_1(k+1) = \Delta t x_2(k) + x_1(k) \tag{3.17}$$

$$x_2(k+1) = x_2(k) + \frac{\Delta t}{M_p}\left\{ A_p[x_3(k) + P_c - P_{atm}] - D_p x_2(k) \right.$$
$$\left. -F_{drag}\left[x_3(k) + P_c, x_2(k)\right] - K_p \times [x_1(k) + x_{p0}]\right\} \tag{3.18}$$

$$x_3(k+1) = \frac{\Delta t \beta}{V_0 + A_p x_1(k)}\left\{ sign\left[u(k) - x_3(k)\right] \times C_d A_{orifice}\sqrt{\frac{2|u(k) - x_3(k)|}{\rho}} - A_p x_2(k)\right\}$$
$$+ x_3(k) \tag{3.19}$$

where Δt refers to the sampling time. For simplicity, define f as the simple representation of the state space model (Equations (3.17) to (3.19)), and define $X(k) = [x_1(k), x_2(k), x_3(k)]^T$.

We define $N = T/\Delta t$ as the number of steps from the initial state to the final state. And the cost function (3.16) becomes

$$g(X) = \sum_{k=0}^{N_1}[x_2(k) - v_m]^2\Delta t + \lambda_1\sum_{k=N_1+1}^{N_2}\left[x_2(k) - \frac{d}{T}\right]^2\Delta t + \lambda_2\sum_{k=N_2+1}^{N-1}[x_2(k) - v_m]^2\Delta t$$
$$+\lambda_3[x_1(N) - d]^2 + \lambda_4[x_2(N) - 0]^2 + \lambda_5\left\{x_3(N) - \left[\frac{K_p(d + x_{p0})}{A_p} + P_{atm} - P_c\right]\right\}^2 \tag{3.20}$$

Consequently, the optimal control problem is to find an optimal control input u to minimize the cost function

$$J(X) = \min_{u \in U} g(X) \tag{3.21}$$

where $X = [X(0), ..., X(N)]$, $u = [u(0), ..., u(N)]$, and U represents the set of feasible control inputs.

3.3.3.3.2 *Optimal Control Using a Customized Dynamic Programming Method*

First, it is desirable to avoid the stiffness and sampling interval constraint associated with the clutch fill dynamic model. One possible way to realize this is to have a proper transformation of the model structure, and therefore change its stiff characteristic. Given the structure of the system model (Equations (3.17) to (3.19)), if $x_1(k + 1)$, $x_2(k + 1)$, $x_3(k + 1)$, and $x_2(k)$ are known, we can calculate the values of $x_1(k)$, $x_3(k)$, and the input $u(k)$ as follows:

$$x_1(k) = x_1(k + 1) - \Delta t x_2(k) = R_1(k) \tag{3.22}$$

$$x_3(k) = R_2(k) \tag{3.23}$$

$$u(k) = sign(W)\left(\frac{W}{C_d A_{orifice}}\right)^2 \frac{\rho}{2} + x_3(k) = R_3(k) \tag{3.24}$$

where

$$W = \left[x_3(k+1) - x_3(k)\right] \times \frac{\left[V_0 + A_p x_1(k)\right]}{\Delta t \beta} + A_p x_2(k)$$

The equation $R_2(k)$ determines the chamber pressure $x_3(k)$ in the step k and involves the drag force F_{drag} term, which has different models depending on the piston velocity x_2. When $x_2(k)$ is not zero, F_{drag} can be obtained from Equation (3.10) and $R_2(k)$ is

$$R_2(k) = \frac{1}{A_p - k_m \times \tanh\left[\dfrac{x_2(k)}{\alpha}\right]} \times \left\{\left[x_2(k+1) - x_2(k)\right] \times \frac{M_p}{\Delta t}\right.$$

$$\left. - A_p(P_c - P_{atm}) + D_p x_2(k) + K_p\left[x_1(k) + x_{p0}\right] + (c_m + k_m P_c)\tanh\left[\frac{x_2(k)}{\alpha}\right]\right\}$$

When $x_2(k)$ is zero but $x_2(k + 1)$ is nonzero, which means that the piston starts moving, the F_{drag} is assumed to be the maximum static friction F_{static} in (3.11) and $R_2(k)$ becomes

$$R_2(k) = \frac{1}{A_p - k_s} \times \left\{\left[x_2(k+1) - x_2(k)\right] \times \frac{M_p}{\Delta t}\right.$$

$$\left. - A_p(P_c - P_{atm}) + D_p x_2(k) + K_p\left[x_1(k) + x_{p0}\right] + (c_s + k_s P_c)\right\}$$

When both $x_2(k)$ and $x_2(k + 1)$ are zero, which means that the piston stays static, the chamber pressure $x_3(k)$ variation is assumed to be small and $R_2(k)$ becomes

$$R_2(k) = x_3(k + 1)$$

For notation simplicity, we can denote Equations (3.22), (3.23), and (3.24) as $[x_1(k), x_3(k), u(k)] = R[x_1(k + 1), x_2(k + 1), x_3(k + 1), x_2(k)]$. Note that as $x_2(k)$ is now predetermined in the inverse dynamic model (Equations (3.22) to (3.24)), it could be regarded as an input, and the new unknown states become $[x_1(k), x_3(k), u(k)]$. Therefore, the Jacobian matrix of (Equations (3.22) to (3.24)) becomes

$$
\frac{\partial [R_1(k), R_2(k), R_3(k)]^T}{\partial [x_1(k+1), x_3(k+1), u(k+1)]^T} =
\begin{bmatrix}
1 & 0 & 0 \\
\dfrac{\partial R_2(k)}{\partial x_1(k+1)} & 0 & 0 \\
\dfrac{\partial R_3(k)}{\partial x_1(k+1)} & \dfrac{\partial R_3(k)}{\partial x_3(k+1)} & 0
\end{bmatrix}
\tag{3.25}
$$

The eigenvalues of (3.25) are always inside or on the unit circle regardless of the value of Δt, which means the dynamic model (Equations (3.22) to (3.24)) is not stiff [14], and therefore the sampling interval constraint can be avoided. Note that the above model transformation still requires the predetermined value for $x_1(k + 1)$, $x_2(k + 1)$, $x_3(k + 1)$, and $x_2(k)$. Interestingly, as the dynamic programming is implemented in a backward fashion, $x_1(k + 1)$, $x_2(k + 1)$, and $x_3(k + 1)$ were calculated in the previous step, and $x_2(k)$ can be discretized into finite grids and predetermined in advance. Note that not all model inversion will result in nonstiff dynamics. For example, the inverse model $X(k) = X(k + 1) - f[X(k + 1), u(k)] \times \Delta t$ (f is the dynamic model (Equations (3.17) to (3.19))) using the backward Euler method is still stiff [15].

Therefore, instead of making combinations of predetermined discrete values of $x_1(k)$, $x_2(k)$, and $x_3(k)$ in the conventional DP [16], we only need to generate the predetermined discrete values for $x_2(k)$ at each step, and then use the values of $x_1(k + 1)$, $x_2(k + 1)$, and $x_3(k + 1)$ from the previous step, together with $x_2(k)$, to obtain the values for $x_1(k)$ and $x_3(k)$. Consequently, unlike the conventional DP, which searches the entire discrete state space [16], the states to search are determined by the system dynamics (Equations (3.22) to (3.24)), and therefore all the discrete states searched in the DP computation are reachable.

However, although the discrete states generated by the above method are all reachable, the number of discrete states will increase from step to step. Suppose the number of discrete states at step $k + 1$ is L_{k+1}, and the number of discrete values of $x_2(k)$ at step k is L, then the total number of discrete states generated for step k would be $L_k = L_{k+1} \times L$. So the number of states will grow very quickly after N steps. To avoid this problem, instead of discretizing the state space into specific discrete values, the customized dynamic programming algorithm divides the state space into several regions. As shown in Figure 3.34 [1], $\{x_2^0(k), x_2^1(k), \ldots, x_2^j(k), \ldots, x_2^L(k)\}$ are the discrete values of x_2 at step k, and $\{X^0(k + 1), X^1(k + 1), \ldots, X^i(k + 1), \ldots, X^{L_{k+1}}(k + 1)\}$ are the possible discrete states calculated from step $k + 1$. In addition, for each discrete value of $x_2(k)$, we can generate several discrete regions on the plane of x_1 and x_3, shown as the black and white blocks in Figure 3.34. Then suppose from

$$
\left[x_1^i(k), x_3^i(k), u^i(k) \right] = R\left[X^i(k + 1), x_2^j(k) \right]
\tag{3.26}
$$

we get $x_1^i(k)$ and $x_3^i(k)$, and suppose that $\left[x_1^i(k), x_2^j(k), x_3^i(k) \right]^T$ lies in region A in Figure 3.34. Subsequently, we can assign the vector $\left[x_1^i(k), x_2^j(k), x_3^i(k) \right]^T$ as the value of region A, and

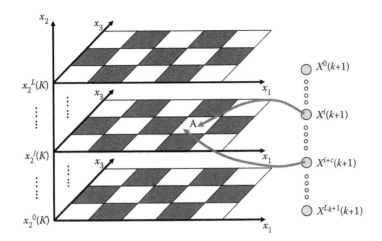

FIGURE 3.34
State space quantization. (From X. Song et al., *Journal of Dynamic Systems, Measurement, and Control*, 133: 054503, 2011. With permission.)

also memorize $\left[x_1^i(k), x_2^j(k), x_3^i(k) \right]^T$ as one of the discrete states for step k. Moreover, we can calculate the cost function based on $\left[x_1^i(k), x_2^j(k), x_3^i(k) \right]^T$ as

$$J_k\left[x_1^i(k), x_2^j(k), x_3^i(k) \right] = \lambda_e \left[x_2^j(k) - v \right]^2 + \hat{J}_{k+1}\left[X^i(k+1) \right] \qquad (3.27)$$

where $\hat{J}_{k+1}$ is the cost function of state $X^i(k + 1)$ at step $k + 1$. Here we also define two symbols v and λ_e, which are equal to v_m in (3.16) and 1, respectively, when step $k \leq N_1$; d/T and λ_1, respectively, when step $k \leq N_2$; and v_m and λ_2, respectively, when $k > N_2$.

In the following calculation, if $\left[x_1^{i+c}(k), x_2^j(k), x_3^{i+c}(k) \right]^T$ obtained from $R\left[X_1^{i+c}(k+1), x_2^j(k) \right]$ also lies in region A, we then calculate the cost function again based on $\left[x_1^{i+c}(k), x_2^j(k), x_3^{i+c}(k) \right]^T$ as

$$J_k\left[x_1^{i+c}(k), x_2^j(k), x_3^{i+c}(k) \right] = \lambda_e \left[x_2^j(k) - v \right]^2 + \hat{J}_{k+1}\left[X^{i+c}(k+1) \right] \qquad (3.28)$$

If $J_k\left[x_1^i(k), x_2^j(k), x_3^i(k) \right]$ is larger than $J_k\left[x_1^{i+c}(k), x_2^j(k), x_3^{i+c}(k) \right]$, we should reassign the value of region A as $\left[x_1^{i+c}(k), x_2^j(k), x_3^{i+c}(k) \right]^T$, and at the same time $\left[x_1^i(k), x_2^j(k), x_3^i(k) \right]^T$ in the discrete space will be replaced by $\left[x_1^{i+c}(k), x_2^j(k), x_3^{i+c}(k) \right]^T$. But if $J_k\left[x_1^i(k), x_2^j(k), x_3^i(k) \right]$ is smaller than $J_k\left[x_1^{i+c}(k), x_2^j(k), x_3^{i+c}(k) \right]$, $\left[x_1^{i+c}(k), x_2^j(k), x_3^{i+c}(k) \right]^T$ will be disregarded and the process goes on.

We summarize the customized dynamic programming algorithm as follows: First, $x_2(k)$ is discretized into a finite grid with size L, and the x_1 and x_3 plane corresponding to each discrete $x_2^j(k)$ is discretized into $L \times L$ regions.

$$\begin{cases} x_2(k) \in \left\{ x_2^1(k), x_2^2(k), ..., x_2^j(k), ..., x_2^L(k) \right\} \\ region_j(k) \in \left\{ reg_j^1(k), reg_j^2(k), ...reg_j^h(k), ..., reg_j^{L \times L}(k) \right\} \end{cases} \qquad (3.29)$$

where *region_j*(k) refers to the discrete regions on the x_1–x_3 plane corresponding to specific $x_2^j(k)$. Also, define λ as a diagonal matrix whose diagonal elements are the weighting factors $λ_3$, $λ_4$, and $λ_5$ in Equation (3.20).

Step $N-1$, for $1 ≤ j ≤ L$:

$$\left[x_1(N-1), x_3(N-1), u(N-1)\right] = R\left[X\_final, x_2^j(N-1)\right] \tag{3.30}$$

$$X^j(N-1) = \left[x_1(N-1), x_2^j(N-1), x_3(N-1)\right] \tag{3.31}$$

$$J_{N-1}[X^j(N-1)] = λ_2[x_2^j(N-1) - v_m]^2$$
$$+ \left\{f[X^j(N-1), u(N-1)] - X_{final}\right\}^T λ\left\{f\left[X^j(N-1), u(N-1)\right] - X_{final}\right\} \tag{3.32}$$

Step k, for $0 ≤ k < N-1$, for $1 ≤ j ≤ L$, and for $1 ≤ i ≤ L_{k+1}$:

$$\left[x_1(k), x_3(k), u(k)\right] = R\left[X^i(k+1), x_2^j(k)\right] \tag{3.33}$$

$$X_{temp} = \left[x_1(k), x_2^j(k), x_3(k)\right] \tag{3.34}$$

If $X_{temp} ∈ reg_j^h(k)$, then

$$J_{k\_temp} = λ_e\left[x_2^j(k) - v\right]^2 + J_{k+1}\left[X^i(k+1)\right] \tag{3.35}$$

If $J_{k\_temp} < J_k[x_2^j(k), reg_j^h(k)]$, then

$$J_k\left[x_2^j(k), reg_j^h(k)\right] = J_{k\_temp} \tag{3.36}$$

$$X_j^h(k) = X_{temp} = \left[x_1(k), x_2^j(k), x_3(k)\right] \tag{3.37}$$

Finally, after we have obtained the minimum cost function for each reachable state, we can easily get the optimal sequence of control input $u = [u(0), …, u(N)]$, which would minimize the total cost function.

In summary, the above algorithm has three advantages over the conventional numerical DP algorithm. First, the system dynamic model is transformed from stiff equations into nonstiff equations. Thus, we can use a smaller number of steps ($N = 800$ for the simulation in Section 3.3.3.4) for DP. Second, all the states in the process are reachable, so by getting states $x_1(k)$ and $x_3(k)$ directly from $x_1(k+1)$, $x_2(k+1)$, $x_3(k+1)$, and $x_2(k)$ using Equation (3.26), $\hat{J}_{k+1}$ in (3.27) can be directly matched with $J_{k+1}[x_1(k+1), x_2(k+1), x_3(k+1)]$, therefore eliminating the approximation errors caused by interpolation in the conventional dynamic programming [16]. Third, as not all the discrete regions need to be considered during the process, the curse of dimensionality [10] problem is mitigated.

3.3.3.4 Simulation and Experimental Results

To validate the optimal clutch fill control, a transmission clutch fixture is designed and built as shown in Figure 3.35 [1].

The hydraulic circuit diagram is shown in Figure 3.36 [1]. A pump with a two-stage pilot-operated relief valve provides the high-pressure fluid. A servo amplifier unit controls the speed of the pump motor. A proportional reducing/relieving valve is used to control the fluid to the clutch chamber.

To measure the motion of the clutch fill process, the clutch system has been instrumented with displacement, pressure, and flow sensors. The clutch piston displacement is measured by two different displacement sensors 180° apart. A Micro Gauging Differential Variable Reluctance Transducer (MGDVRT) is mounted on one location of the clutch piston, and a laser sensor is mounted 180° apart. Due to the rubber sealing and manufacturing tolerance, the friction around the circular piston is not necessarily balanced. This may cause the piston to twist around the shaft while moving. Therefore, it is necessary to measure the piston motion at different locations simultaneously and calculate the average piston displacement. In addition, the clutch system input pressure is measured using an Omega pressure sensor PX209-030G5V with measurement range from 0 to 30 psi and resolution of 0.075 psi, and a Max machinery flow meter G015 with a range from 0.15 to 15 lpm, and the time constant of 1.7 ms is used to measure the input flow rate.

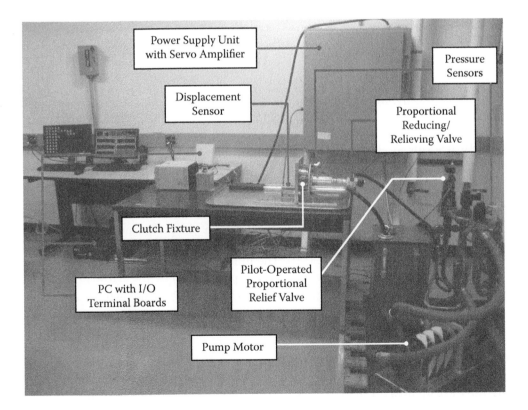

FIGURE 3.35
Clutch fill experimental setup. (From X. Song et al., *Journal of Dynamic Systems, Measurement, and Control*, 133: 054503, 2011. With permission.)

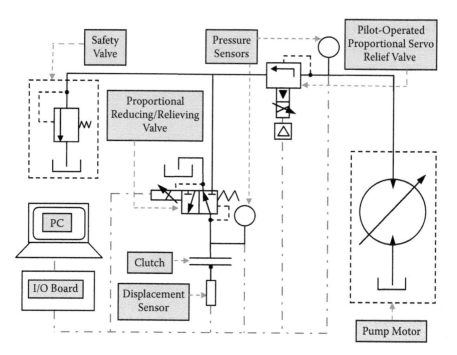

FIGURE 3.36
The hydraulic circuit scheme diagram. (From X. Song et al., *Journal of Dynamic Systems, Measurement, and Control,* 133: 054503, 2011. With permission.)

3.3.3.4.1 System Identification

To implement the dynamic programming and optimal control, the parameters of the clutch system model (Equations (3.7) to (3.9)) need to be identified. The effective mass of the piston M_p, the piston surface area A_p, the piston return spring constant K_p, the preload of the return spring x_{p0}, the stick friction peak value F_{static}, the orifice area $A_{orifice}$, the fluid density ρ, the discharge coefficient C_d, the bulk modulus β, and the clutch chamber volume V_0 can be measured or obtained directly. As the clutch is not rotating in the lab experimental setup, the centrifugal pressure P_c is zero.

The F_{static}, which is the maximum drag force when the clutch stays static, is measured by recording the clutch chamber pressure at the start of the piston motion. In the experiment, the clutch piston is kept static at the specific travel distance and then moves to the next position as shown in Figure 3.37 [1]. The minimum pressure P_{min} required to move the piston at the specified position is obtained as pointed out by the arrow in Figure 3.37. Therefore, F_{static} in the corresponding chamber pressure can be calculated by

$$F_{static} = P_{min} \times A_p - K_p \times (x_1 + x_{p0}) \tag{3.38}$$

The remaining model parameters, the damping coefficient D_p, the piston seal damping coefficient α, and the piston seal drag force F_{drag}, while the piston is moving, are identified using the least-squares estimation approach [17]. The identified model is then compared with the experimental data using the input pressure profile shown in Figure 3.38(a) [1], and the modeling error is shown in Figure 3.38(b) [1]. Finally, the measured and identified parameter values are presented in Table 3.2 [1].

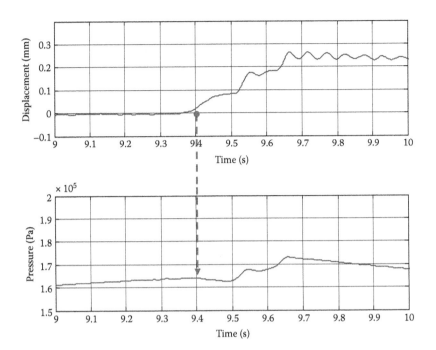

FIGURE 3.37
Experiments for measuring the stick friction F_{static}. (From X. Song et al., *Journal of Dynamic Systems, Measurement, and Control*, 133: 054503, 2011. With permission.)

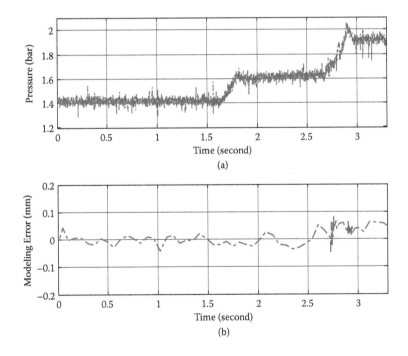

FIGURE 3.38
System identification model verification. (a) Input pressure profile. (b) Modeling error. (From X. Song et al., *Journal of Dynamic Systems, Measurement, and Control*, 133: 054503, 2011. With permission.)

TABLE 3.2

Parameter Values of the System Dynamic Model

M_p	0.4 (kg)	T	0.3 (s)
x_{p0}	1.5928 (mm)	Δt	0.000375 (s)
K_p	242,640 (N/m)	$A_{orifice}$	2.0442e-5 (m²)
D_p	135.4 (N/m/s)	ρ	880 (kg/m³)
A_p	0.00628 (m²)	C_d	0.7
α	4.1054e-6 (m/s)	P_{atm}	1 (bar)
V_0	7.8e-5 (m³)	β	1625 (bar)
k_m	0.001517 (m²)	k_s	0.00153 (m²)
c_m	5.22 (N)	c_s	5.26 (N)

Source: X. Song et al., *Journal of Dynamic Systems, Measurement, and Control*, 133: 054503, 2011. With permission.

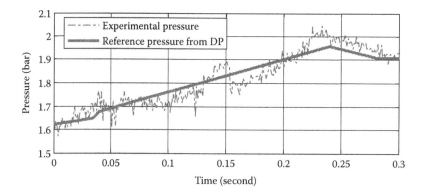

FIGURE 3.39
Optimal input pressure and the experimental tracking results. (From X. Song et al., *Journal of Dynamic Systems, Measurement, and Control*, 133: 054503, 2011. With permission.)

3.3.3.4.2 Clutch Fill Simulation and Experimental Results

In this section, both simulation and experimental results are reported to validate the proposed control method. An optimal input pressure is derived to achieve the desired clutch fill velocity profile using the customized dynamic programming method. The desired final state conditions are $x_1(T) = d = 0.000725$ (m), $x_2(T) = 0$ (m/s), and $x_3(T) = 1.91 \times 10^5$ (Pa). Figure 3.39 [1] (solid line) shows the optimal control input. The total steps for DP equal 800. The number of discrete values for $x_2(k)$ at each step is 100, and the number of discrete regions for each discrete $x_2(k)$ value is 100 × 100. For the customized DP method, we only evaluate the states that can be reached. The computation time for the customized DP method is 22 min with a 1.86 GHZ computer.

The resulting optimal control input pressure is then implemented in the experiment to verify its performance as shown in Figure 3.39 (dashed line). Due to the short time duration (0.3 s) and total displacement (0.725 mm), precise pressure and motion control of the clutch fill process is very challenging. In addition, the clutch mechanism is extremely sensitive to pressure rise in the clutch chamber when the clutch piston starts to move, which adds up to the intricacy of controlling the clutch piston motion. This is because the current clutch design behaves as an on/off switch during the clutch fill process, where the input pressure required to start the piston motion is only 0.3 bar lower than the final pressure (see Figure 3.39). Despite these challenges, the clutch piston displacement, velocity, and input flow profiles from the experiments exhibit optimal shape, as shown in Figure 3.40 [1]. The experimental velocity profile in Figure 3.40(b) is obtained by numerical differentiation of the displacement data, and its optimal shape can be further verified by the smooth input flow rate shown in Figure 3.40(c).

The repeatability issue is also verified using multiple experiments with the same desired pressure inputs shown in Figure 3.39. The experimental results in Figure 3.41 exhibit good repeatability. Specifically, Figure 3.41(a) presents the piston displacements from five groups of experiments, which clearly show that all the displacement trajectories have similar optimal shapes and can overlap each other. Figure 3.41(b) shows the error histogram of all the data points in the five trajectories compared with the desired one, which again exhibits good repeatability.

We also added ±5% and ±10% perturbations to the dynamic model parameters (D_p, α, β, F_{drag}, ρ, V_o, $A_{orifice}$) and simulated the system performance using the designed control input in Figure 3.39. The trajectories shown in Figure 3.42 [1] arrived in acceptable intervals

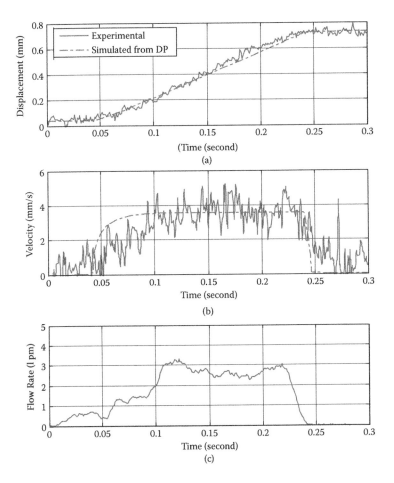

FIGURE 3.40
Experimental results for clutch displacement, velocity, and input flow rate. (From X. Song et al., *Journal of Dynamic Systems, Measurement, and Control*, 133: 054503, 2011. With permission.)

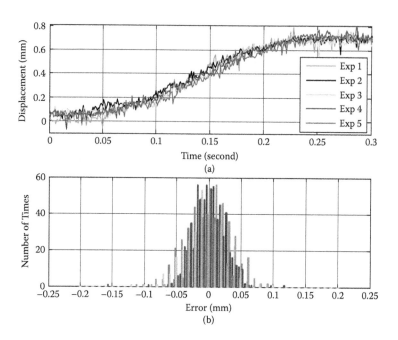

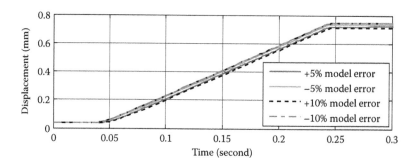

FIGURE 3.41

Clutch fill repeatability test. (a) Five groups of displacement profiles. (b) Histogram of data error compared with optimal trajectory. (From X. Song et al., *Journal of Dynamic Systems, Measurement, and Control*, 133: 054503, 2011. With permission.)

FIGURE 3.42

Clutch fill robustness test. (From X. Song et al., *Journal of Dynamic Systems, Measurement, and Control*, 133: 054503, 2011. With permission.)

around the desired final states and thus demonstrated system robustness against those model parameter variations. In addition, random perturbations within ±10% interval are also added to those model parameters, and the resulting piston final displacements at the end of the clutch fill are collected through 100 tests. The clutch fill final piston position errors are then projected into a histogram shown in Figure 3.43 [1]. The histogram shows that most of the clutch trajectories finally arrive within a small interval around 0.725 mm, which is the desired displacement of the clutch fill.

The influence of the time delay caused by the solenoid valve was explored as well. The desired optimal clutch fill piston displacement is compared to that with solenoid valve delay in Figure 3.44 [1]. Because of the unique velocity profile proposed in Figure 3.33, the

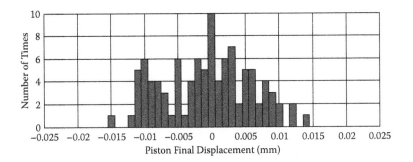

FIGURE 3.43
Histogram of clutch fill piston final position. (From X. Song et al., *Journal of Dynamic Systems, Measurement, and Control*, 133: 054503, 2011. With permission.)

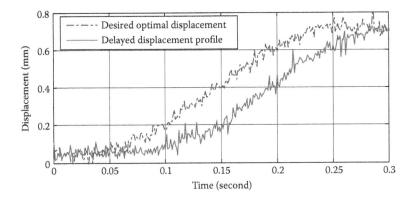

FIGURE 3.44
Experimental data demonstrating clutch fill robustness on time delay. (From X. Song et al., *Journal of Dynamic Systems, Measurement, and Control*, 133: 054503, 2011. With permission.)

clutch fill with solenoid delay can still reach the final position within the required clutch fill duration (0.3 s).

Comparison with a nonoptimal clutch fill process is also shown in Figure 3.45 [1]. It can be seen that the nonoptimal approach results in a high clutch piston velocity spike, which represents the high peak flow demand that could only be met with a larger pump.

3.3.4 Pressure-Based Clutch Feedback Control

In this section, we will study the pressure-based feedback control for transmission clutches [18]. This will include both clutch fill and clutch engagement control. Currently there is no pressure sensor inside the clutch chamber, and therefore the clutch control in the automatic transmissions is either in an open-loop fashion or controlled using speed signal as the feedback. However, the speed signal can only be used as the feedback variable during the clutch engagement, but not for the clutch fill. With the increasing demand of transmission efficiency and performance, a more precise clutch shift control is necessary, which calls for a more effective closed-loop clutch control. Therefore, a sensor that can measure the clutch motion needs to be installed to close the loop. There are two possible ways of measuring the clutch motion: measuring the clutch piston displacement or

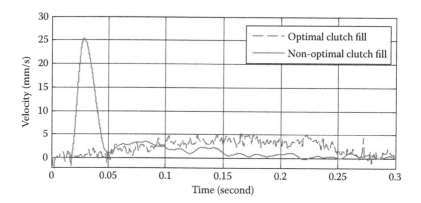

FIGURE 3.45
Optimal and nonoptimal clutch fill velocities profile comparison. (From X. Song et al., *Journal of Dynamic Systems, Measurement, and Control*, 133: 054503, 2011. With permission.)

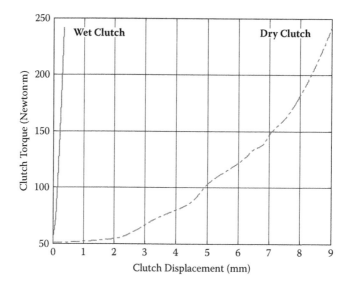

FIGURE 3.46
Characteristic curves for dry and wet clutches. (From X. Song and Z. Sun, *IEEE/ASME Transactions on Mechatronics*, 17(3): 534–546, 2012. With permission.)

the pressure inside the clutch chamber. In this section, the pressure feedback is selected for "wet" clutch control for three reasons. First, although the transferred torque can be calculated from the clutch pack displacement, the displacement vs. torque curve for wet clutch is very steep compared with "dry" clutch [7, 19], as shown in Figure 3.46 [18], which makes it difficult to obtain an accurate torque estimate from the displacement. In contrast, the pressure-based information can be directly related to the transferred torque during the clutch engagement. Second, with a compact transmission design, it is very difficult to package a displacement sensor that moves with the clutch piston. Third, for the wet clutch used in automatic transmissions, the total displacement of the clutch piston is about 1~2 mm, and a high-resolution displacement sensor for such a small range is usually cost-prohibitive for mass production.

3.3.4.1 System Dynamics Modeling

Figure 3.47 shows a simplified schematic diagram of the clutch actuation system. The main components include a pump, hydraulic control valves, a clutch assembly, pressure, and displacement sensors. The proportional pressure-reducing valve controls the flow in and out of the clutch chamber. When the clutch fill begins, the valve will connect the clutch chamber to the high-pressure source, which is regulated by a relief valve, and the high-pressure fluid flows into the chamber and pushes the piston toward the clutch packs. Once the clutch fill ends, the valve will control the chamber pressure to further increase until the clutch packs are fully engaged. When the clutch is disengaged, the proportional pressure-reducing valve connects the clutch chamber to the tank, and the piston return spring pushes the clutch piston back to the disengaged position. The clutch dynamic model consists of the valve dynamics, the clutch mechanical dynamics, and the clutch chamber pressure dynamics, which will be presented in the following sections.

The valve used to control the clutch chamber pressure is a three-way two-position proportional pressure-reducing valve, as shown in Figure 3.48(a) [18]. Port 1 is connected to the clutch chamber with pressure P_r, port 2 is connected to the supply pressure P_{high}, and port 3 is connected to the tank. The orifice between the clutch chamber (port 1) and the supply pressure (port 2) is determined by the spool position, which is controlled by the input voltage.

When there is no control voltage, the spool is kept at the top position by the return spring. At this position port 3 is connected to port 1. Therefore, the fluid inside the clutch chamber

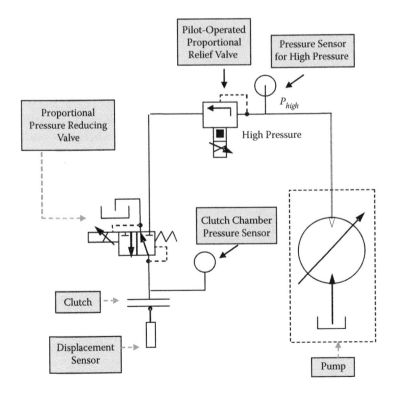

FIGURE 3.47

Hydraulic circuit diagram of the clutch actuation system. (From X. Song and Z. Sun, *IEEE/ASME Transactions on Mechatronics*, 17(3): 534–546, 2012. With permission.)

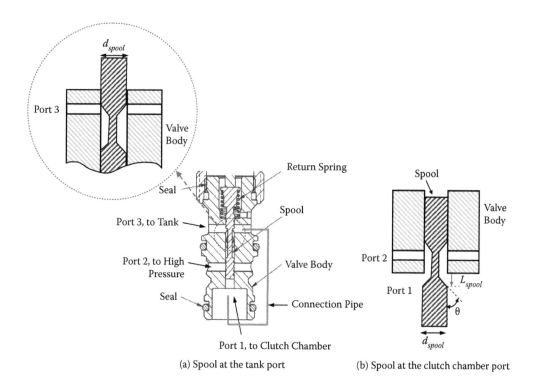

(a) Spool at the tank port   (b) Spool at the clutch chamber port

FIGURE 3.48
Cross-sectional view of the proportional pressure-reducing valve. (From X. Song and Z. Sun, *IEEE/ASME Transactions on Mechatronics*, 17(3): 534–546, 2012. With permission.)

will flow to the tank. When a positive voltage is exerted on the magnetic coil, the induced magnetic force will push the spool toward port 1. As the spool connects port 1 to port 2, the high-pressure fluid will flow in to the clutch chamber. The increased pressure in port 1 will push the spool upward and eventually close the orifice between port 1 and port 2. Clearly, the spool position is determined by the magnetic force F_{mag}, the returning spring force F_{spring}, and the chamber pressure P_r. The spool dynamics can be described as [18]

$$\dot{L}_{spool} = v_{spool}$$

$$\dot{v}_{spool} = \frac{1}{M_{spool}} \Big[F_{mag}(Vol) - K_{spring}(L_{spool} + L_{pre-load})$$

$$- D_{spool} v_{spool} - A_{spool} P_r \Big]$$

(3.39)

where

$$F_{mag}(Vol) = K_f \times i = K_f \times \left(\frac{i_{max} - i_{min}}{Vol_{max}} Vol + i_{min} \right)$$

(3.40)

L_{spool} is the spool position, v_{spool} is the spool velocity, M_{spool} is the spool mass, K_{spring} is the spool spring constant, A_{spool} is the cross-sectional area of the spool, and D_{spool} is the damping coefficient. $L_{preload}$ is the spool position where the orifice between port 1 and port 2 is just closed. F_{mag} is the magnetic force, which is determined by the coil magnetic constant K_f

and the current i. The current i is generated by the power amplifier and can be calculated using the input voltage *Vol*. i_{max} and i_{min} are the maximum and minimum currents that can be generated from the power amplifier, and Vol_{max} is the maximum control voltage corresponding to i_{max}.

As shown in Figure 3.48(b) [18], given the spool position L_{spool}, the orifice area $A_{orifice}$ between high-pressure port 2 and chamber pressure port 1 is

$$A_{orifice}(L_{spool}) = \pi L_{spool} \sin(\theta) \times (d_{spool} - L_{spool} \sin\theta \cos\theta) \tag{3.41}$$

where d_{spool} is the diameter of the spool, and θ is the spool surface slant angle.

Similarly, as shown in Figure 3.48(a), the orifice area A_{dump} between tank port 3 and pressure port 1 is

$$A_{dump}(L_{spool}) = -\pi L_{spool} \sin(\theta) \times (d_{spool} + L_{spool} \sin\theta \cos\theta) \tag{3.42}$$

Then the flow dynamics across the valve orifice is

$$Q(L_{spool}, P_r) = \begin{cases} sign(P_h - P_r)C_d \sqrt{\dfrac{2|P_h - P_r|}{\rho}} A_{orifice}(L_{spool}), & L_{spool} > 0 \\ 0, & L_{spool} = 0 \\ -C_d \sqrt{\dfrac{2|P_r|}{\rho}} A_{dump}(L_{spool}), & L_{spool} < 0 \end{cases} \tag{3.43}$$

where ρ is the fluid density, C_d is the discharge coefficient, $A_{orifice}$ is the orifice area connecting the clutch chamber (port 1) and the supply pressure (port 2), and A_{dump} is the orifice area connecting the clutch chamber (port 1) and the tank (port 3). When the spool position $L_{spool} > 0$, the clutch chamber is connected to the high pressure. When the spool position $L_{spool} < 0$, the clutch chamber is connected to the tank.

The clutch mechanical system model and clutch chamber pressure dynamics model can be found in Section 3.3.3.1.

3.3.4.1.1 Overall System Dynamic Model

The overall clutch system dynamic model including the valve dynamics, the mechanical dynamics, and the chamber pressure dynamics can be summarized as follows [18]:

$$\dot{L}_{spool} = v_{spool}$$

$$\dot{v}_{spool} = \frac{1}{M_{spool}} \Big[F_{mag}(Vol) - K_{spring}(L_{spool} + L_{pre-load}) \tag{3.44}$$

$$-D_{spool} v_{spool} - A_{spool} P_r \Big]$$

$$\dot{P}_r = \frac{\beta(P_r)}{V} \Big[Q(L_{spool}, P_r) - A_p x_2 \Big] \tag{3.45}$$

$$\dot{x}_1 = x_2 \tag{3.46}$$

$$\dot{x}_2 = \frac{1}{M_p} \times \left[A_p \times (P_r + P_c - P_{atm}) - D_p x_2 \right.$$
$$\left. - F_{drag}(P_r + P_c, x_2) - F_{res}(x_1 + x_{p0}) \right] \tag{3.47}$$

F_{res} is the displacement-dependent resistance force. During the clutch fill, the resistance force comes from the piston return spring; thus, the force F_{res} depends on the spring stiffness constant K_{cs}. During the clutch engagement, the resistance force is due to the squeezing of the clutch packs, and therefore the resistance force F_{res} can be modeled as

$$F_{res} = \begin{cases} K_{cs} \times (x_1 + x_{p0}) & (x_1 \leq x_{fill}) \\ F_{en}(x_1 + x_{p0}) & (x_1 > x_{fill}) \end{cases} \tag{3.48}$$

where the function $F_{en}(x_1 + x_{p0})$ includes both spring force and the nonlinear clutch pack reaction force; x_{fill} is the clutch piston position at the end of the clutch fill.

F_{drag} is the piston seal drag force, which is dependent on the piston motion.

$$F_{drag} = \begin{cases} \left[k_m(P_r + P_c) + c_m \right] \times sign(x_2) & (x_2 \neq 0) \\ F_{stick} & (x_2 = 0) \end{cases} \tag{3.49}$$

where k_m and c_m are constant, and F_{stick} is the static stick friction force from Kanopp's stick-slip model [11]. Further detail of the mechanical dynamic model can be found in Section 3.3.3.1.

Further investigation reveals that the time constant of the pressure-reducing valve is far less than the time constant of the clutch actuation system pressure dynamics, which means that the dynamic behavior of the reducing valve can be neglected. Therefore, to simplify the control design, the proportional valve spool dynamics is converted into a static mapping;

$$L_{spool} = \frac{1}{K_{spring}} \left[F_{mag}(u) - K_{spring} \times L_{pre-load} - A_{spool} P_r \right] \tag{3.50}$$

Here, we denote $u = Vol$ as the control input to the proportional valve. Therefore, the system dynamics (Equations (3.44) and (3.45)) becomes

$$\dot{P}_r = \frac{\beta(P_r)}{V} \left[Q(u, P_r) - A_p x_2 \right] \tag{3.51}$$

where

$$Q(u, P_r) = \begin{cases} sign(P_h - P_r) C_d \sqrt{\dfrac{2|P_h - P_r|}{\rho}} A_{orifice}(L_{spool}), & L_{spool}(u, P_r) > 0 \\ 0, & L_{spool}(u, P_r) = 0 \\ -C_d \sqrt{\dfrac{2|P_r|}{\rho}} A_{dump}(L_{spool}), & L_{spool}(u, P_r) < 0 \end{cases} \tag{3.52}$$

Finally, the reduced order system dynamics includes Equations (3.46) to (3.52).

In addition, for a given flow rate q, the required control input u can also be determined based on Equations (3.41), (3.42), (3.50), and (3.52). This relationship will be used in the next section, and the mapping from q to u can be written as

$$U(q, P_r) = \frac{Vol_{max}}{i_{max} - i_{min}} \times \left[\frac{(L_{spool}K_{spring} + K_{spring}L_{pre-load} + A_{spool}P_r)}{K_f} - i_{min} \right] \tag{3.53}$$

where

$$L_{spool} = \begin{cases} \dfrac{d_{spool} - \sqrt{d_{spool}^2 - \dfrac{4q\cos(\theta)}{\pi sign(P_h - P_r)C_d\sqrt{\dfrac{2|P_h - P_r|}{\rho}}}}}{\sin(2\theta)}, & q > 0 \\[4ex] 0, & q = 0 \\[2ex] \dfrac{-d_{spool} + \sqrt{d_{spool}^2 + \dfrac{4q\cos(\theta)}{\pi C_d\sqrt{\dfrac{2|P_r|}{\rho}}}}}{\sin(2\theta)}, & q < 0 \end{cases} \tag{3.54}$$

Equation (3.54) is to calculate the spool position L_{spool} corresponding to a specific valve opening orifice based on Equations (3.41) and (3.42). The relationship between Equations (3.52) and (3.53) can be written as $Q\{U(q, P_r), P_r\} = q$.

3.3.4.2 Robust Nonlinear Controller and Observer Design

3.3.4.2.1 Sliding Mode Controller Design for Pressure Control

The clutch system (Equations (3.46) to (3.52)) is a third-order nonlinear system. As both the control input u (the valve voltage) and the control output P_r (the chamber pressure) appear in the dynamic equation (3.51), the system has a nonlinear dynamics with relative degree 1 [20]. Equations (3.46) and (3.47) are the system internal dynamics, the equilibrium point of which is asymptotically stable. Therefore, the nonlinear system is minimum phase [20]. Note that the internal dynamics in this application is a spring mass damper system; therefore, the minimum phase feature also suggests that the whole system will be stabilized as long as the states other than the internal dynamics states could be stabilized. This unique feature suggests applying the sliding mode controller, which can ensure system robustness with a relatively low order controller design [20].

Define the tracking error e_2 as the difference between the desired pressure trajectory r and the actual measurement P_r.

$$e_2 = P_r - r \tag{3.55}$$

And define another error term, e_1, the derivative of which is equal to e_2.

$$\dot{e}_1 = e_2 \tag{3.56}$$

With the pressure dynamics in (3.51), we have

$$\dot{e}_2 = \dot{P}_r - \dot{r}$$

$$= \frac{\beta(P_r)}{V}\left[Q(u, P_r) + \Delta_2(u, P_r) - A_p x_2\right] + \Delta_1(P_r) - \dot{r} \qquad (3.57)$$

$$= \frac{\beta(P_r)}{V}Q(u, P_r) - \frac{\beta(P_r)}{V}A_p x_2 - \dot{r} + \Delta_1(P_r) + \frac{\beta(P_r)}{V}\Delta_2(u, P_r)$$

where $\Delta_1(P_r)$ represents the model uncertainty of the pressure dynamics (3.51), and $\Delta_2(u, P_r)$ represents the model uncertainty of the pressure-reducing valve flow dynamics (3.52). Bounds of the uncertainty terms will be obtained experimentally in a later section.

Define the sliding surface S as

$$S = k_1 e_1 + e_2 \qquad (3.58)$$

where k_1 is a weighting parameter.

Then the controller can be designed as

$$u = -U\left\{\frac{V}{\beta(P_r)} \times \left[k_1 e_2 - \frac{\beta(P_r)A_p}{V}x_2 - \dot{r}\right] - v_u, P_r\right\} \qquad (3.59)$$

where U is defined in Equation (3.53); v_u is a controller term to be designed later. As the piston velocity x_2 cannot be measured directly due to the lack of a displacement and velocity sensor, an observer is needed to estimate x_1 and x_2. The estimated states are not only fed back to the controller (3.59), but also used to evaluate the piston displacement, and therefore the clutch fill status. The observer design for piston motion will be presented in the next section. With the observed x_2, the control input becomes

$$u = -U\left\{\frac{V}{\beta(P_r)} \times \left[k_1 e_2 - \frac{\beta(P_r)A_p}{V}\hat{x}_2 - \dot{r}\right] - v_u, P_r\right\}$$

$$= -U\left\{\frac{V}{\beta(P_r)} \times \left[k_1 e_2 - \frac{\beta(P_r)A_p}{V} \times (x_2 + \Delta_3(\hat{x}_2)) - \dot{r}\right] - v_u, P_r\right\} \qquad (3.60)$$

where $\hat{x}_2$ is the estimate of x_2 and $\Delta_3(\hat{x}_2)$ is the estimation error.

Then the sliding surface becomes

$$\dot{S} = k_1 \dot{e}_1 + \dot{e}_2$$

$$= k_1 e_2 + \dot{e}_2 \qquad (3.61)$$

By substituting Equation (3.57) and Equation (3.60) into Equation (3.61), it becomes

$$\dot{S} = \frac{\beta(P_r)}{V}v_u + \Delta(u, P_r, \hat{x}_2) \qquad (3.62)$$

where $\Delta(u, P_r, \hat{x}_2) = \Delta_1(P_r) + \frac{\beta(P_r)}{V}\Delta_2(u, P_r) + \frac{\beta(P_r)A_p}{V}\Delta_3(\hat{x}_2).$

With $V_{lp} = (1/2)S^2$ as the Lyapunov function for Equation (3.62), we have

$$\dot{V}_{lp} = S\dot{S} = \frac{\beta(P_r)}{V} v_u S + \Delta S$$

$$\leq \frac{\beta(P_r)}{V} v_u S + |\Delta| \times |S|$$

(3.63)

if

$$\left| \frac{\Delta \times V}{\beta(P_r)} \right| \leq \psi(P_r, \hat{x}_2) + \lambda_0 |v_u|$$

(3.64)

where $\psi(P_r, \hat{x}_2)$ is a positive continuous function, and λ_0 is a real number in the $(0,1)$ interval. Then v_u can be designed as

$$v_u = -\gamma(P_r, \hat{x}_2) sign(S)$$

(3.65)

where $\gamma(P_r, \hat{x}_2)$ is the controller gain, and design $\gamma(P_r, \hat{x}_2) \geq \dfrac{\psi(P_r, \hat{x}_2)}{1 - \lambda_0} + \gamma_0$, $\gamma_0 > 0$.

Substituting Equations (3.64) and (3.65) into Equation (3.63) gives [20]

$$\dot{V}_{lp} \leq \frac{\beta(P_r)}{V} v_u S + |\Delta| \times |S|$$

$$\leq \frac{\beta(P_r)}{V} \times \left[-\gamma(P_r, \hat{x}_2) + \psi(P_r, \hat{x}_2) + \lambda_0 \gamma(P_r, \hat{x}_2) \right] |S|$$

$$\leq -\frac{\beta(P_r)}{V} \times \gamma_0 \times (1 - \lambda_0) |S|$$

$$< 0$$

which proves the convergence of the controller design.

Finally, the controller is given as follows:

$$u = -U \left\{ \frac{V}{\beta(P_r)} \times \left[k_1 e_2 - \frac{\beta(P_r) A_p}{V} \times \hat{x}_2 - \dot{r} \right] + \gamma(P_r, \hat{x}_2) sign(S), P_r \right\}$$

(3.66)

In the real-time implementation, to suppress the chattering problem, the *sign* function in (3.66) is approximated by a saturation function as [20]

$$sign(S) = sat\left(\frac{S}{\sigma} \right)$$

where σ is a constant positive number. The control structure can therefore be summarized in Figure 3.49 [18].

As shown in Equation (3.66), implementation of the controller requires the uncertainty bound and the velocity estimation. A conservative uncertainty bound will make the controller gain γ too large, and thus result in chattering. Therefore, a proper estimation of the model uncertainty is critical. In addition, the piston velocity estimate $\hat{x}_2$ is actually a compensation term in the controller design (3.66), and it will influence the pressure tracking performance, especially during the clutch fill, where the piston velocity is higher. Besides, the clutch fill status can also be evaluated based on the piston motion estimate.

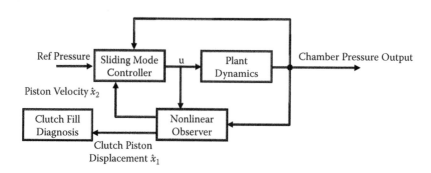

FIGURE 3.49
Controller design structure. (From X. Song and Z. Sun, *IEEE/ASME Transactions on Mechatronics*, 17(3): 534–546, 2012. With permission.)

Therefore, a piston motion observer is necessary and will be presented in the following section.

3.3.4.2.2 Observer Design

The observer is designed to estimate the clutch piston displacement x_1 and velocity x_2 using the pressure measurement P_r. During the clutch fill process, the piston velocity x_2 is nonzero and the drag force F_{drag} (Equation (3.49)) is only a function of the pressure. Also, the resistance force F_{res} (Equation (3.48)) is the piston return spring force, and thus is linear as well. Suppose y is the measurement of the chamber pressure P_r. The observer for the clutch fill process is designed as [18]:

$$\dot{\hat{x}}_1 = \hat{x}_2 + L_1 \times \left(\frac{V}{\beta(y)} \dot{y} - Q(u,y) + A_p \hat{x}_2 \right)$$

$$\dot{\hat{x}}_2 = \frac{1}{M_p} \left[A_p \times (y + P_c - P_{atm}) - D_p \hat{x}_2 - F_{drag}(y + P_c) - K_{cs} \times (\hat{x}_1 + x_{p0}) \right] \qquad (3.67)$$

$$+ L_2 \times \left(\frac{V}{\beta(y)} \dot{y} - Q(u,y) + A_p \hat{x}_2 \right)$$

Suppose the estimation error is denoted as

$$\varepsilon_1 = x_1 - \hat{x}_1$$

$$\varepsilon_2 = x_2 - \hat{x}_2 \qquad (3.68)$$

Combining Equations (3.67) and (3.46) to (3.51), we have

$$\dot{\varepsilon}_1 = \varepsilon_2 - L_1 \left(\frac{V}{\beta(y)} \dot{y} - Q(u,y) + A_p \hat{x}_2 \right)$$

$$= \varepsilon_2 - L_1(-A_p x_2 + A_p \hat{x}_2)$$

$$= \varepsilon_2 + L_1 A_p \varepsilon_2 \qquad (3.69)$$

$$\dot{\varepsilon}_2 = \frac{1}{M_p} \left[-D_p \varepsilon_2 - K_{cs} \varepsilon_1 \right] + L_2 A_p \varepsilon_2$$

from which we can get

$$
\begin{bmatrix} \dot{\varepsilon}_1 \\ \dot{\varepsilon}_2 \end{bmatrix} = \begin{bmatrix} 0 & 1 + L_1 A_p \\ -\dfrac{K_{cs}}{M_p} & \left(L_2 A_p - \dfrac{D_p}{M_p} \right) \end{bmatrix} \begin{bmatrix} \varepsilon_1 \\ \varepsilon_2 \end{bmatrix} \tag{3.70}
$$

The estimation error will converge to zero if the observer gains L_1 and L_2 are selected so that the eigenvalues of the error dynamics (Equation (3.70)) is on the left half plane. Besides, the piston initial position might differ from the initial state of the observer, so the observer gain should be designed to enable fast error convergence as well.

During the clutch engagement, the resistance force F_{res} (Equation (3.48)) is nonlinear. However, the experimental calibration reveals that the wet clutch characteristic curve can be approximated with two straight lines. Therefore, the observer for clutch engagement can adopt the same structure as (3.67):

$$
\dot{\hat{x}}_1 = \hat{x}_2 + L_1 \times \left(\frac{V}{\beta(y)} \dot{y} - Q(u, y) + A_p \hat{x}_2 \right)
$$

$$
\dot{\hat{x}}_2 = \frac{1}{M_p} \Big[A_p \times (y + P_c - P_{atm}) - D_p \hat{x}_2 - F_{drag}(y + P_c) \tag{3.71}
$$

$$
- K_{en}(\hat{x}_1) \times (\hat{x}_1 + x_{p0}) \Big] + L_2 \times \left(\frac{V}{\beta(y)} \dot{y} - Q(u, y) + A_p \hat{x}_2 \right)
$$

where K_{en} is the stiffness parameter characterizing the clutch characteristic curve approximated by two straight lines.

Finally, as the observer design includes the derivative of the pressure measurement, in practice, the values of L_1 and L_2 are constrained to avoid the amplification of the measurement noise.

3.3.4.2.3 Experimental Results

In this section, experimental results are presented to validate the proposed control methods. A picture of the experimental setup is shown in Figure 3.50 [18]. The sliding mode controller together with the observer is implemented on the clutch fixture. The continuous time controller is converted into a discrete controller with a sampling rate of 1 ms.

The observer estimation result is shown in Figure 3.51(a) [18] compared with the average displacement measured by the displacement sensors. As we only have pressure measurement in the real-time feedback control, the clutch piston displacement information is not available directly. Then the observer displacement estimate could be used to diagnose the clutch fill status. For example, the clutch fill should finish within 250 ms and the piston should travel up to 0.7 mm at the end of the clutch fill. If the observer estimation at the end of clutch fill is not close to the desired value, then further pressure control action is expected to amend the clutch fill status. In practice, due to the linear approximation of the nonlinear clutch characteristic curve and the clutch pack wear, the mechanical model uncertainty during the clutch engagement is typically larger compared to that during the clutch fill. A practical approach to maintain accurate estimation is to assign a larger L_2 gain (Equation (3.71)) during the clutch engagement, so that the pressure dynamics

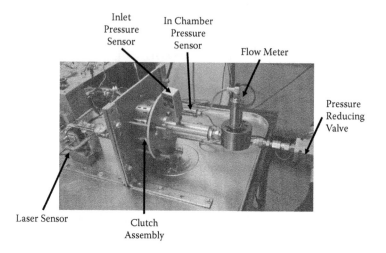

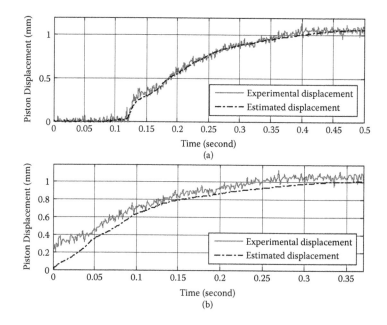

FIGURE 3.50

The experimental setup for pressure-based clutch actuation. (Only pressure is used in the real-time feedback, and other sensors are installed for dynamic modeling purposes.) (From X. Song and Z. Sun, *IEEE/ASME Transactions on Mechatronics*, 17(3): 534–546, 2012. With permission.)

FIGURE 3.51

The nonlinear observer estimation results. (a) Estimation with accurate initial state. (b) Estimation with inaccurate initial state. (From X. Song and Z. Sun, *IEEE/ASME Transactions on Mechatronics*, 17(3): 534–546, 2012. With permission.)

can dominate in the observer estimation. The effect of the initial state of the observer is shown in Figure 3.51(b) [18]. Even with a 0.25 mm initial clutch position estimation error, the estimate converges to the actual value quickly. The effect of model uncertainties is investigated as well. Figure 3.52 [18] shows the estimation results when the piston return spring parameter K_p, the spring preload x_{p0}, the clutch damping coefficient D_p, the piston

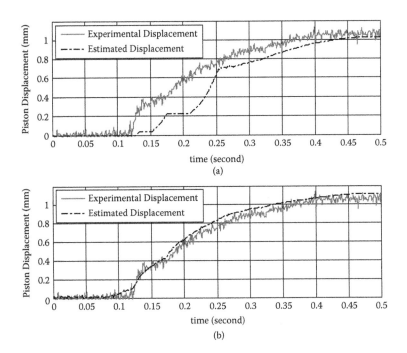

FIGURE 3.52
The estimation with clutch fill mechanical dynamics perturbation. (a) Estimation with low L_2 gain in clutch fill. (b) Estimation with high L_2 gain in clutch fill. (From X. Song and Z. Sun, *IEEE/ASME Transactions on Mechatronics*, 17(3): 534–546, 2012. With permission.)

drag force coefficients k_m and c_m, and the oil density are all perturbed by 5%. With a small L_2 gain for the clutch fill observer, the weighting on the pressure measurement is not enough to overcome the mechanical model error, and thus the estimation discrepancy during the clutch fill phase (from 0.1 to 0.3 s) is evident as shown in Figure 3.52(a). The estimation can then converge during the clutch engagement due to a larger L_2 gain for the clutch engagement observer. Figure 3.52(b) [18] shows the estimation result if a larger L_2 gain is assigned for the clutch fill observver. However, in practice, the value of the L_2 gain is constrained. First, as the observer design has a derivative term, large observer gain may amplify the measurement noise. Second, increasing the gain L_2 will diminish the effect of the mechanical dynamics, which can prohibit the error convergence if the initial state error exists.

Figure 3.53 [18] shows the pressure tracking result for clutch fill and clutch engagement. The initial pressure is at 1.68 bar, which is the critical pressure counteracting the spring preload. When the clutch fill starts, the pressure increases to 1.97 bar and then drops down to 1.9 bar. This pressure profile design is to enable the optimal clutch fill [4, 21]. During the clutch engagement, the pressure quickly rises to 7 bar to squeeze the clutch packs. The pressure notch during engagement is to simulate the clutch slipping control. The control signal is shown in Figure 3.54(a) [18], and the flow in rate is shown in Figure 3.54(b) [18]. The first peak flow rate corresponds to the clutch fill phase, and the second one corresponds to the clutch engagement. During the clutch fill, the piston velocity is faster than that during the clutch engagement. Therefore, the flow-out rate induced by the piston motion during the clutch fill is bigger, which needs more inflow to increase the pressure.

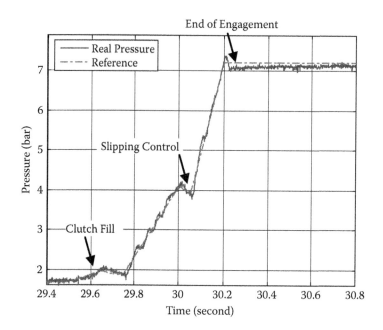

FIGURE 3.53
Pressure tracking for clutch fill and clutch engagement. (From X. Song and Z. Sun, *IEEE/ASME Transactions on Mechatronics*, 17(3): 534–546, 2012. With permission.)

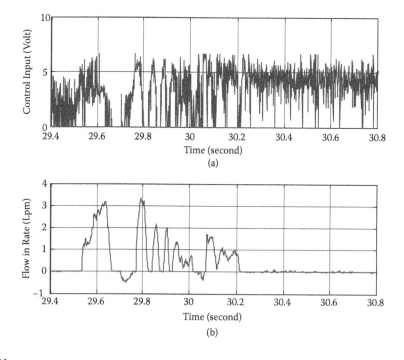

FIGURE 3.54
Control input and the flow-in rate for clutch control. (a) Pressure control input for clutch fill and clutch engagement. (b) Flow rate for clutch fill and clutch engagement control. (From X. Song and Z. Sun, *IEEE/ASME Transactions on Mechatronics*, 17(3): 534–546, 2012. With permission.)

3.4 Driveline Dynamics and Control

A representative diagram for vehicle driveline is shown in Figure 3.55. Typical starting devices for the transmission include torque converter, manually controlled clutch, and automatically controlled clutch. Typical disturbances for the driveline include engine firing pulse, road condition, and driver input: tip in or tip out, etc. The objective for driveline control is to avoid driveline vibration and improve fuel efficiency.

A more detailed schematic for the vehicle driveline is shown in Figure 3.56. Based on this figure, a dynamic model for the driveline is shown as

$$
\begin{cases}
J_e \dot{w}_e = T_e - T_c \\[2mm]
\left[J_c + J_{eq}(i_g \cdot i_d) \right] \cdot \dot{w}_c = T_c - \dfrac{1}{i_g \cdot i_d} \left[k\Delta\theta + \beta \left(\dfrac{w_c}{i_g \cdot i_d} - w_w \right) \right] \\[2mm]
J_w \dot{w}_w = k\Delta\theta + \beta \left(\dfrac{w_c}{i_g \cdot i_d} - w_w \right) - T_L \\[2mm]
\dot{\Delta\theta} = \dfrac{w_c}{i_g \cdot i_d} - w_w
\end{cases}
\tag{3.72}
$$

where w_e, w_c, and w_w are the angular velocity of the engine, clutch, and wheel; J_e, J_c, and J_w are the moment of inertia of the engine, clutch, and wheel; T_e is the engine torque input; T_c is the torque transmitted through the clutch; T_L is the load torque at the wheel; i_g and i_d are the ratio of the gear and differential; J_{eq} is the equivalent moment of inertia of the whole gear box; k is the spring constant; β is the damping coefficient; and $\Delta\theta$ is the relative angle between the gear box and wheel.

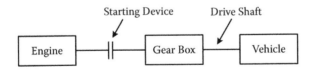

FIGURE 3.55
Block diagram for vehicle driveline.

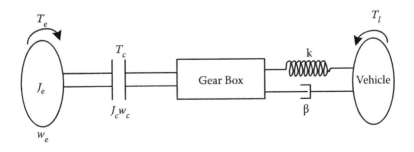

FIGURE 3.56
Schematic diagram for vehicle driveline.

When the clutch is engaged, that is, $w_e = w_c$, we have

$$\left[J_e + J_c + J_{eq}(i_g \cdot i_d) \right] \dot{w}_c = T_e - \frac{1}{i_g \cdot i_d} \left[k\Delta\theta + \beta \left(\frac{w_c}{i_g \cdot i_d} - w_w \right) \right]$$

$$J_w \dot{w}_w = k\Delta\theta + \beta \left(\frac{w_c}{i_g \cdot i_d} - w_w \right) - T_L$$

$$\Delta\dot{\theta} = \frac{w_c}{i_g \cdot i_d} - w_w$$

Since $w_e = w_c$, the system becomes third order. From the dynamic models, we can see that the clutch torque T_c is critical for system performance and fuel economy. If T_c is not controlled properly, we could have engine flare, engine misfire, driveline vibration, and sluggish gear shift.

Let's further examine the controllability of the system. First let's further simplify the system dynamics: assume $\Delta\theta = 0$; i.e., the driveshaft is a solid connection. We then have

$$J_e \dot{w}_e = T_e - T_c$$

$$J_v \dot{w}_c = T_c - T_l$$

Let $x_1 = \theta_e$, $x_2 = w_e$, $x_3 = \theta_c$, $x_4 = w_c$; we have

$$\dot{x}_1 = x_2$$

$$\dot{x}_2 = \frac{1}{J_e}[T_e - T_c]$$

$$\dot{x}_3 = x_4$$

$$\dot{x}_4 = \frac{1}{J_v}[T_c - T_l]$$

$$\begin{bmatrix} \dot{x}_1 \\ \dot{x}_2 \\ \dot{x}_3 \\ \dot{x}_4 \end{bmatrix} = \begin{bmatrix} 0 & 1 & 0 & 0 \\ 0 & 0 & 0 & 0 \\ 0 & 0 & 0 & 1 \\ 0 & 0 & 0 & 0 \end{bmatrix} \begin{bmatrix} x_1 \\ x_2 \\ x_3 \\ x_4 \end{bmatrix} + \begin{bmatrix} 0 \\ \frac{1}{J_e}[T_e - T_c] \\ 0 \\ \frac{1}{J_v}[T_c - T_l] \end{bmatrix}$$

If the only control variable is T_c, we have

$$A = \begin{bmatrix} 0 & 1 & 0 & 0 \\ 0 & 0 & 0 & 0 \\ 0 & 0 & 0 & 1 \\ 0 & 0 & 0 & 0 \end{bmatrix}, B = \begin{bmatrix} 0 \\ -\dfrac{1}{J_e} \\ 0 \\ \dfrac{1}{J_v} \end{bmatrix}$$

And the controllability matrix is

$$
\begin{bmatrix}
0 & 0 & -\dfrac{1}{J_e} & 0 \\
0 & 0 & 0 & -\dfrac{1}{J_e} \\
0 & 0 & \dfrac{1}{J_v} & 0 \\
0 & 0 & 0 & \dfrac{1}{J_v}
\end{bmatrix}
$$

which is not full rank.

This suggests that by controlling the clutch torque only, we cannot control the engine speed and the vehicle speed independently. However, if we have two control variables, T_c and T_e, we then have

$$
A = \begin{bmatrix}
0 & 1 & 0 & 0 \\
0 & 0 & 0 & 0 \\
0 & 0 & 0 & 1 \\
0 & 0 & 0 & 0
\end{bmatrix}, \quad
B = \begin{bmatrix}
0 & 0 \\
-\dfrac{1}{J_e} & \dfrac{1}{J_e} \\
0 & 0 \\
\dfrac{1}{J_v} & 0
\end{bmatrix}
$$

The controllability matrix is then

$$
\begin{bmatrix}
0 & 0 & 0 & 0 & -\dfrac{1}{J_e} & \dfrac{1}{J_e} & 0 & 0 \\
0 & 0 & 0 & 0 & 0 & 0 & -\dfrac{1}{J_e} & \dfrac{1}{J_e} \\
0 & 0 & 0 & 0 & \dfrac{1}{J_v} & 0 & 0 & 0 \\
0 & 0 & 0 & 0 & 0 & 0 & \dfrac{1}{J_v} & 0
\end{bmatrix}
$$

which is full rank.

This suggests that to have complete control over the engine and vehicle speed, two control inputs, the engine torque and the clutch torque, are required. However, the engine torque is often controlled by the drive through the throttle. The clutch torque, however, can be fully controlled by the driveline controller. So it would be interesting to understand the exact impact of the clutch torque.

Define $x_5 = x_1 - x_3$, $x_6 = x_2 - x_4$. By considering the clutch slip speed as the state variable, and using T_c as the only control, we have

$$
\begin{bmatrix} \dot{x}_5 \\ \dot{x}_6 \end{bmatrix} = \begin{bmatrix} 0 & 1 \\ 0 & 0 \end{bmatrix} \begin{bmatrix} x_5 \\ x_6 \end{bmatrix} + \begin{bmatrix} 0 \\ \dfrac{1}{J_e}T_e + \dfrac{1}{J_v}T_l - \left(\dfrac{1}{J_e} + \dfrac{1}{J_v}\right)T_c \end{bmatrix}
$$

So

$$A = \begin{bmatrix} 0 & 1 \\ 0 & 0 \end{bmatrix}, \ B = \begin{bmatrix} 0 \\ -\left(\dfrac{1}{J_e} + \dfrac{1}{J_v}\right) \end{bmatrix}$$

The controllability matrix is

$$\begin{bmatrix} -\left(\dfrac{1}{J_e} + \dfrac{1}{J_e}\right) & 0 \\ 0 & -\left(\dfrac{1}{J_e} + \dfrac{1}{J_e}\right) \end{bmatrix}$$

which is full rank.

This suggests that the clutch torque can have complete control over the slip speed. This is important when designing clutch control to achieve the desired driveline objective.

When simulating driveline dynamics, various driving cycles are often used, such as the Federal Testing Procedure (FTP) used by the Environmental Protection Agency (EPA). Given a specific driving cycle, the vehicle power demand can be calculated using the method presented in Chapter 1. The engine operating point and the transmission gear ratio can also be calculated. This will allow the systematic evaluation of different driveline control methodologies in a realistic driving scenario.

References

1. X. Song, A. Mohd Zulkefli, Z. Sun, and H. Miao, Automotive Transmission Clutch Fill Control Using a Customized Dynamic Programming Method, *Journal of Dynamic Systems, Measurement, and Control*, 133: 054503, 2011.
2. Z. Sun and H. Kumar, Challenges and Opportunities in Automotive Transmission Control, presented at Proceedings of American Control Conference, Portland, OR, June 8–10, 2005.
3. X. Song, C. Wu, and Z. Sun, Design, Modeling, and Control of a Novel Automotive Transmission Clutch Actuation System, *IEEE/ASME Transactions on Mechatronics*, 17(3): 582–587, 2012.
4. H. Miao, Z. Sun, J. Fair, J. Lehrmann, and S. Harbin, Modeling and Analysis of the Hydraulic System for Oil Budget in an Automotive Transmission, presented at Proceedings of ASME Dynamic Systems and Control Conference, Ann Arbor, MI, October 20–22, 2008.
5. M. Montanari, F. Ronchi, C. Rossi, A. Tilli, and A. Tonielli, Control and Performance Evaluation of a Clutch Servo System with Hydraulic Actuation, *Journal of Control Engineering Practice*, 12(11): 1369–1379, 2004.
6. J. Horn, J. Bamberger, P. Michau, and S. Pindl, Flatness-Based Clutch Control for Automated Manual Transmissions, *Journal of Control Engineering Practice*, 11(12): 1353–1359, 2003.
7. L. Glielmo, L. Iannelli, and V. Vacca, Gearshift Control for Automated Manual Transmissions, *IEEE/ASME Transactions on Mechatronics*, 11(1): 17–26, 2006.
8. R.E. Bellman, *Dynamic Programming*, Princeton, NJ: Princeton University Press, 1957.
9. R.E. Bellman and S.E. Dreyfus, *Applied Dynamic Programming*, Princeton, NJ: Princeton University Press, 1962.

10. A.P. De Madrid, S. Dormido, and F. Morilla, Reduction of the Dimensionality of Dynamic Programming: A Case Study, in *Proceedings of American Control Conference*, San Diego, CA, June 1999, pp. 2852–2856.
11. D. Karnopp, Computer Simulation of Stick-Slip Friction in Mechanical Dynamic Systems, *Journal of Dynamic Systems, Measurement, and Control*, 107(1): 100–103, 1985.
12. X. Song, A. Zulkefli, Z. Sun, and H. Miao, Modeling, Analysis, and Optimal Design of the Automotive Transmission Ball Capsule System, *Journal of Dynamic Systems, Measurement, and Control*, 132: 021003, 2010.
13. X. Song, Z. Sun, X. Yang, and G. Zhu, Modeling, Control and Hardware-in-the-Loop Simulation of an Automated Manual Transmission, *Journal of Automobile Engineering*, 224(2): 143–160, 2010.
14. E. Hairer and G. Wanner, *Solving Ordinary Differential Equations II*, 2nd ed., Springer Series in Computational Mathematics, Berlin: Springer, 1996.
15. W. Gautschi, *Numerical Analysis: An Introduction*, Boston: Birkhauser, 1997.
16. J.M. Kang, I. Kolmanovsky, and J.W. Grizzle, Dynamic Optimization of Lean Burn Engine Aftertreatment, *Journal of Dynamic Systems, Measurement, and Control*, 123(2): 153–160, 2001.
17. J.L. Crassidis and J.L. Junkins, *Optimal Estimation of Dynamic Systems*, CRC Applied Mathematics and Nonlinear Science Series, Boca Raton, FL: Chapman and Hall/CRC Express LLC, 2004.
18. X. Song and Z. Sun, Pressure Based Clutch Control for Automotive Transmissions Using a Sliding Mode Controller, *IEEE/ASME Transactions on Mechatronics*, 17(3): 534–546, 2012.
19. F. Vasca, L. Iannelli, A. Senatore, and G. Reale, Torque Transmissibility Assessment for Automotive Dry-Clutch Engagement, *IEEE/ASME Transactions on Mechatronics*, 16(3): 564–573, 2011.
20. H. Khalil, *Nonlinear Systems*, 3rd ed., Upper Saddle River, NJ: Prentice Hall, 2002.
21. X. Song, A. Mohd Zulkefli, and Z. Sun, Automotive Transmission Clutch Fill Optimal Control: An Experimental Investigation, presented at Proceedings of American Control Conference, Baltimore, MD, June 30–July 2, 2010.

4

Design, Modeling, and Control of Hybrid Systems

4.1 Introduction to Hybrid Vehicles

As shown in Chapter 1, the fundamental challenge that limits powertrain efficiency is the varying vehicle speed and power demand in real time and the dependence of engine efficiency on speed and load conditions. Powertrain hybridization is one of the most effective ways for addressing this challenge. As the name suggests, a hybrid powertrain consists of the conventional internal combustion engine (ICE) and an alternative power source. Therefore, a hybrid transmission is needed to combine the power from the ICE and the alternative power source and send it to the wheels of the vehicle. Depending on the type of alternative power source and the architecture of the hybrid transmission, hybrid vehicles can be divided into different categories. Detailed discussions on the types of hybrid vehicles and the hybrid architectures will be provided in Sections 4.1.1 and 4.2.

The basic fuel-saving mechanisms for hybrid vehicles include: (1) optimize the engine operating condition independent from the vehicle operating condition using the extra degree of freedom enabled by the alternative power source, (2) use an energy storage element to provide a buffer between the engine power generation and the vehicle power demand, and (3) provide regenerative braking by converting the vehicle kinetic energy during braking into energy storage for future use. For the first fuel-saving mechanism, a large energy storage element is not necessary. However, for the second and third fuel-saving mechanisms, an energy or power storage element is essential to achieve the desired functions, which has been a key technical challenge for hybrid vehicle development.

The research and development of hybrid vehicles has been mainly focused on the design of the hybrid architecture [1] and the energy management of the hybrid powertrain [2, 3]. In this chapter, we will first present different types of hybrid vehicles, and then discuss the hybrid architecture design, and finally show the hybrid powertrain dynamics and control.

4.1.1 Various Types of Hybrid Vehicles

So far several different types of hybrid vehicles have been proposed: hybrid electric vehicle (HEV), hydraulic hybrid vehicle, pneumatic hybrid vehicle, and hybrid mechanical vehicle. As the names suggest, the main difference among the various hybrid vehicles is the type of alternative power source employed.

The hybrid electric vehicle complements the ICE with the electrical motor/generator and uses a battery to store the energy. The hybrid hybrid vehicle uses the hydraulic accumulator to store energy or power and the hydraulic pump/motor as the actuation device. The pneumatic hybrid vehicle uses a compressed air tank to store energy and the pneumatic pump/motor to transmit power. The hybrid mechanical vehicle stores energy with a mechanical fly wheel and the mechanical energy can be transmitted with clutches and other mechanisms, such as gears.

Each type of hybrid vehicle has its advantages and disadvantages. The main properties for differentiating the hybrid vehicles include energy density, power density, cost, reliability, and ease for control. So far, the hybrid electric vehicle has the largest market share and is mainly focused on passenger cars. The hydraulic hybrid vehicle has been the main hybrid technology for heavy vehicles. As will be shown in Section 4.2, the hybrid architecture is applicable to different hybrid vehicles even though the underlying physical systems are different. A similar statement can be made for hybrid powertrain dynamics and control, which will be presented in Section 4.3.

4.2 Hybrid Architecture Analysis

4.2.1 Parallel Hybrid Architecture

As shown in Figure 4.1, the parallel architecture employs an alternative power unit (APU) in parallel with the ICE. The APU can either assist the torque output from the engine or load the engine. This will allow the engine to operate at loading conditions that are different from the vehicle power demand (see Chapter 1) to optimize its efficiency and emissions. However, with the parallel architecture, the speed of the ICE cannot be independently varied from the vehicle speed since it is still constrained by the transmission (see Chapter 3). This will greatly limit the flexibility of the parallel architecture. Besides changing the loading conditions of the engine, the parallel architecture can also enable the stop-and-go function and provide limited regenerative braking. The stop-and-go function is realized by shutting down the engine when the vehicle comes to a full stop and restarting the engine using the APU when the vehicle is ready to launch again. This function will help to reduce the engine idling time and save fuel. The regenerative braking can be achieved by converting the vehicle kinetic energy into the alternative energy (such as electricity) through the APU (such as a generator). But this function is often limited by the size of the APU and the energy storage device, as well as safety considerations of the braking operation. Of course, regenerative braking is also dependent on the amount of kinetic energy available when braking is initiated. This can vary drastically depending on the size

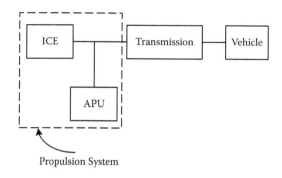

Propulsion System

FIGURE 4.1
Diagram of parallel hybrid architecture.

of the vehicle and its application. For compact passenger cars, the amount of kinetic energy ready for regenerative braking is limited.

Typical examples of the parallel architecture include the belt alternator system (BAS), the integrated motor assist (IMA) system, the hydraulic launch assist system, etc. The first two examples are electric hybrid vehicles, while the third example is a hydraulic hybrid vehicle. A key advantage of the parallel architecture is that it is compatible with the conventional powertrain system and is relatively easy to implement. It also offers redundancy protection for system reliability.

4.2.2 Series Hybrid Architecture

As shown in Figure 4.2, the series hybrid architecture converts the engine output into the alternative energy (such as electricity) first and then converts it back to mechanical energy to propel the vehicle using the APU (such as a motor). This architecture allows the engine to operate at speed and load conditions that are independent from the vehicle power demand provided that the energy storage device (such as a battery) is large enough. If no energy storage device is provided or the capacity of the energy storage device is too small to cover the power demand of the vehicle, the series architecture will work as a continuously variable transmission (CVT) where the engine output power has to match the vehicle power demand, but the specific engine operating point (speed and load) can still be optimized for the given power output. Compared with the parallel architecture, the series architecture provides significant flexibility for optimizing the engine operation, and therefore improves its fuel efficiency and reduces emissions. However the drawback is that there is double conversion of energy in the series architecture, namely, from mechanical energy to electrical energy, and then back to mechanical energy. Such double conversion decreases the power transmission efficiency. So the overall efficiency (engine efficiency and the power transmission efficiency) has to be evaluated carefully.

Unlike the parallel architecture, the series architecture has seen fewer applications on the market. This is mainly due to the fact that it is quite different from the conventional powertrain and requires a large energy storage device. This situation may change as more efficient engines and components are developed. An interesting example under this architecture is the free piston engine (FPE) where the internal combustion engine and the electrical generator or hydraulic pump are integrated into one device. The FPE doesn't output mechanical power; instead, it converts fuel into electrical power or fluid power. Without the crankshaft, the FPE is a modular device that can offer infinite variable compression ratio control. This can enable advanced combustions such as homogeneous charge compression ignition (HCCI), which offers significantly higher thermal efficiency and less emissions.

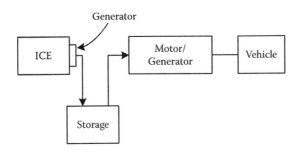

FIGURE 4.2
Diagram of the series hybrid architecture.

4.2.3 Power-Split Hybrid Architecture

As shown in Figure 4.3, the power-split architecture employs a hybrid transmission to combine the power from the ICE and the motor/generator sets and transmit it to the vehicle. The hybrid transmission often consists of one or several planetary gear sets. A key property of the planetary gear set is that for the speeds at the three nodes (sun, carrier, and ring), only after two speeds are determined can the third speed be decided (for details see Chapter 3). This is essentially a two degrees of freedom system, which enables the flexibility of varying engine operating conditions (speed and torque) by controlling the motor/generator set. The power-split architecture combines the benefits of the parallel and series architecture by splitting the engine power into the mechanical and the electrical path. Specifically, the mechanical path refers to the power flowing through the planetary gear sets and transmitting to the vehicle mechanically. The electrical path refers to the power that is converted from mechanical power into electrical power through a generator, and eventually converted back to mechanical power through a motor. This architecture enables the engine to vary its operating condition independent of the vehicle operating condition, while still transmitting part of its power through the mechanical path to avoid double conversion of energy. This architecture offers a good compromise between flexibility and power transmission efficiency.

There are several vehicles on the market that employ the power-split architecture. One example is the THS system, as shown in Figure 4.4. The engine is connected to the carrier, and the generator is connected to the sun gear. Ring gear is connected to the output. The motor is coupled to the output shaft. For any given vehicle speed, the engine speed can be

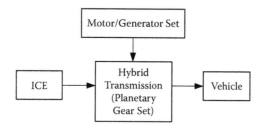

FIGURE 4.3
Diagram of the power-split architecture.

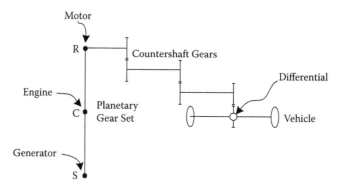

FIGURE 4.4
Diagram of the THS system.

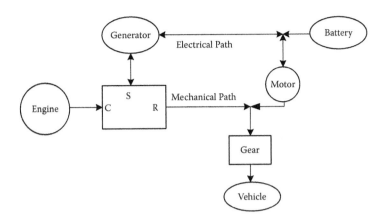

FIGURE 4.5
Power flow diagram.

varied to the optimal operating point by adjusting the speed of the generator. For this reason, the power-split hybrid transmission is also referred to as the electrical continuously variable transmission (ECVT). As shown in Figure 4.5, the power-split hybrid architecture can be controlled to transmit the power through both the mechanical path and the electrical path. The ratio between the power transmissions in these two paths can be controlled in real time.

4.3 Hybrid System Dynamics and Control

4.3.1 Dynamic Models for Hybrid System

The dynamics of the hybrid system is more complex than the dynamics of the conventional powertrain system due to the alternative power source and the hybrid architecture. For the parallel or series hybrid, the model can be obtained by combining the models of the individual elements. But for the power-split hybrid, the model is different due to the extra degree of freedom introduced by the hybrid transmission.

As shown in Figure 4.6, the speeds of the engine, the generator, and the motor satisfy the following relationship:

$$W_g S + W_r R = W_e (R + S) \tag{4.1}$$

Figure 4.7 shows the free body diagram of the power-split hybrid. We assume the pinion is massless; that is why the force on the ring gear is the same as the force on the sun gear.

$$(I_G + I_S)\dot{W}_G = F \cdot S - T_G \tag{4.2}$$

$$(I_E + I_C)\dot{W}_e = T_e - F \cdot R - F \cdot S \tag{4.3}$$

$$\left(I_R + I_M + \frac{mR_{tire}^2}{k^2}\right)\dot{W}_R = (T_M + FR) - \frac{1}{k}(T_f + mgf_r R_{tire} + f_d V^2 R_{tire}) \tag{4.4}$$

where T_f is the brake torque, f_r is the friction coefficient, and f_d is the wind resistance coefficient.

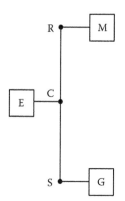

FIGURE 4.6
Block diagram of the power-split hybrid architecture.

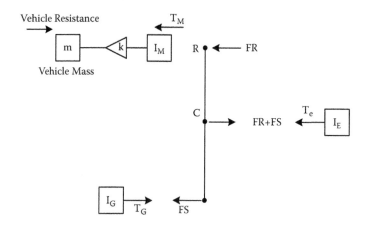

FIGURE 4.7
Free body diagram of the power-split architecture.

From (4.1), we can get $\dot{W}_G S + \dot{W}_R R = \dot{W}_e (R + S)$.

The state of the charge for the battery can be modeled as

$$SOC = -\frac{I_{batt}}{Q_{max}} \tag{4.5}$$

where Q_{max} is the battery capacity and I_{batt} is the current drawing from the battery.

The battery power is

$$P_{batt} = V_{oc} I_{batt} - I_{batt}^2 R_{batt} \tag{4.6}$$

where V_{OC} is the battery open-circuit voltage and R_{batt} is the battery internal resistance.

Since the battery power is either used to drive the motor or charged through the generator, we also have

$$P_{batt} = (T_G W_G \eta_G^K \eta_{i1}^K + T_M W_M \eta_M^K \eta_{i2}^K) \tag{4.7}$$

where η_G is the efficiency of the generator, η_M is the efficiency of the motor, η_{i1} is the efficiency of inverter 1, η_{i2} is the efficiency of inverter 2, $K = 1$ is the charging (power goes to the battery), and $K = -1$ is the discharging (power comes out of the battery).

From (4.5) to (4.7), we get

$$\dot{SOC} = -\frac{V_{OC} - \sqrt{V_{OC}^2 - 4(T_G W_G \eta_G^K \eta_{i1}^K + T_M W_M \eta_M^K \eta_{i2}^K)R_{batt}}}{2R_{batt}Q_{max}} \tag{4.8}$$

This model can be further extended by including the driveline dynamics, motor/generator, and power electronics dynamics.

The dynamic model of the permanent magnet (PM) AC synchronous machine is

$$L_d \dot{I}_d = V_d - R_S I_d + \eta_P L_q W_m I_q$$

$$L_q \dot{I}_q = V_q - R_S I_q - \eta_P L_d W_m I_d - \psi_m \eta_P W_m \tag{4.9}$$

$$T_M = \frac{3}{2} n_P [\psi_m I_q + (L_d - L_q)I_d I_q]$$

where V_d, V_q are the direct and quadrature axis voltages, I_d, I_q are the direct and quadrature axis currents, L_d, L_q are the direct and quadrature axis inductances, W_m is the motor speed, T_M is the motor torque, R_S is the series phase resistance, n_P is the number of pole pairs, and ψ_m is the permanent magnet flux.

4.3.2 Hybrid System Control

The majority of the published work on hybrid system control is focused on energy management [4–12]. In this section we will discuss two energy management strategies (Sections 4.3.2.1 and 4.3.2.2) and the driveline dynamics control of the hybrid system (Section 4.3.2.3).

4.3.2.1 Transient Emission and Fuel Efficiency Optimal Control

Both gasoline engines and diesel engines can be used for the hybrid application. The diesel HEV is expected to offer even more fuel savings given the high efficiency of the diesel engine. However, the diesel HEV has an inherent disadvantage of high NOx and soot emissions, especially under transient operations [13, 14]. Targeted at achieving both the global energy optimization and transient emissions control, a two-mode hybrid powertrain energy management strategy is designed [15]. First, dynamic programming (DP) is employed to ensure the global fuel efficiency optimization and battery state of charge (SOC) sustainability for any given driving cycles. On this basis, during selected emissions-reducing modes, the management strategy locally modifies the engine operations into another trajectory with the purpose of reducing the high transient emissions, and more importantly, at the end of every emissions-reducing mode, the locally optimized engine operating trajectory will be driven back to match the globally optimized trajectory generated by the DP for the succeeding operation. This design ensures the global optimization of the fuel efficiency (albeit with some slight deviations in some areas) and battery SOC sustainability since the DP-optimized engine/battery operations are recovered after every local emission optimization. Specifically, in the fuel efficiency-improving mode, the management strategy makes use of the DP algorithm to seek the global optimization of the fuel efficiency over the entire driving cycle; while in the emission-reducing mode, this strategy utilizes a linear quadratic regulator (LQR) to locally optimize the operating trajectory in a short time horizon, to suppress the surging emissions due to the sudden engine torque transients. Here, it is worth noting that we avoid directly adding the transient emission optimization task into the DP global optimization algorithm, which will otherwise significantly increase the computational burden of DP.

4.3.2.1.1 *Control-Oriented Emission Model*

Targeted at realizing simultaneous fuel consumption and emission control, both the engine fuel consumption and emission models are needed. Different from the static mapping-based fuel consumption model, the emission model must involve higher-order transient dynamics. Ideally, to precisely describe the complex thermodynamics and chemistry of the engine combustion, the complete engine transient emissions model should contain as many high-fidelity physics-based submodels as possible that can catch the various engine processes (in-cylinder processes, inlet/exhaust manifold operation, exhaust gas recirculation (EGR) operation, turbocharger operation, and so on) and identify the effects of all the key operating parameters (engine speed, fuel injection mass, inlet air pressure, exhaust gas temperature, EGR opening, and so on). However, to implement the transient emission control in real time, a simplified, control-oriented model with well-selected input variables is required.

Various control-oriented emission models have been reported. A dynamic model composed of steady-state emission and transient emission correction was adopted in [16], where some process variables that represent the engine combustion or cylinder charge characteristics are used as inputs to characterize the transient emission effects. Otherwise, actual EGR opening rate, fresh air mass flow, and fuel/air ratio are also commonly used for transient correction [17, 18]. However, the previous emission models are commonly used for improving the engine operation within the engine control unit (ECU), but not for the vehicle-level energy management strategy. Recently, some studies on the emission models that are suitable for the vehicle-level control are also reported. In [19], a Volterra series-based model with fuel flow rate and engine speed as the inputs is presented.

Aiming at developing a hybrid vehicle energy management strategy that can optimize the fuel consumption over the entire driving cycle and reduce local transient emissions, a data-driven emission model that is capable of predicting the transient emission is built and identified by experiments. The modeling and validation process consists of two steps:

1. Input parameters selection. Although empirical data have shown that the engine internal variables (inlet air pressure, exhaust gas temperature, EGR opening, etc.) are closely coupled with both NOx and soot emissions, including those variables will increase the dimension of the emission model and, consequently, the computational burden of the vehicle-level energy management. Therefore, the proposed model chooses the engine external variables (driver's control inputs)—engine speed ω_e and engine torque T_e—as the inputs. Such selection is based on the observation that the other key variables (such as the inlet air pressure, exhaust temperature, and so on) are ultimately determined by the current engine speed and torque command (or their derivatives).

   As a result, this input parameters selection brings us two advantages. First, the engine torque and speed are more accessible than the engine internal variables (i.e., ECU control actions), and second, these control inputs can directly determine the engine fuel consumption in the static mapping model, which is advantageous for coordinating the optimal controls of both the fuel consumption and emissions in the energy management strategy. The experimental data of a diesel engine (John Deere 4045HF, Tier IV, 4 cylinders, 4.5 L) demonstrate the steady-state relationship between the engine fuel consumption/emissions and engine torque/ speed, as shown in Figure 4.8(a–c) [25], where the engine loads correspond to different engine torques.

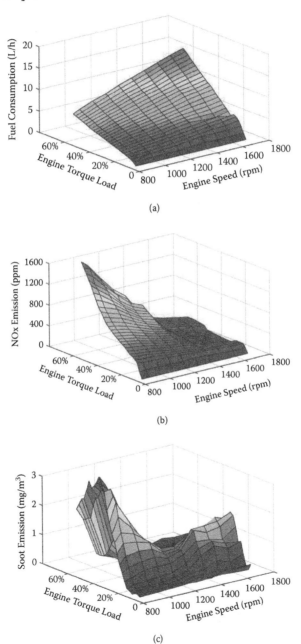

(a)

(b)

(c)

FIGURE 4.8
Steady-state fuel consumption and emissions maps. (a) Fuel consumption. (b) NOx emission. (c) Soot emission. (From Y. Wang et al., *Journal of Automobile Engineering*, 227(11): 1546–1561, 2013.)

In addition, to capture the transient emission dynamics without introducing new independent variables, the derivatives of engine speed $\dot{\omega}_e$ and engine torque $\dot{T}_e$ are used as additional inputs. The physical explanation of $\dot{\omega}_e$ and $\dot{T}_e$ as additional inputs to capture the transient emission dynamics is illustrated in Figures 4.9 and 4.10 [25]. Under different engine load transient conditions (different torque derivatives), the fuel injection flow rate will present different transients, which in turn affects the transient air/fuel ratio, and hence the emission output. The faster the engine load changes, i.e., the larger the torque derivative $\dot{T}_e$, the higher the transient fuel flow rate becomes, as shown in Figure 4.9. However, the inlet airflow rate is limited by the air pressure dynamics. Eventually, the transient air/fuel ratio that can directly cause undesired emission output is coupled with the torque derivative $\dot{T}_e$, as shown in Figure 4.10.

2. Control-oriented model development and validation. With the well-selected input variables, a proper mathematic model should be constructed to describe the nonlinear emission dynamics with relatively compact structure. To capture the nonlinear

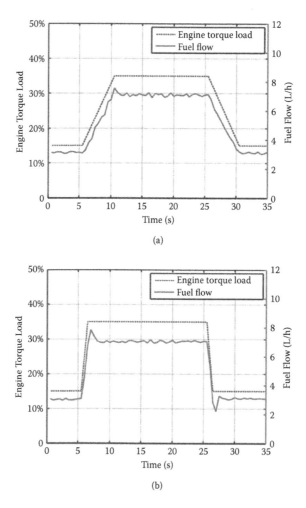

FIGURE 4.9
Engine fuel injections under different torque transients. (a) Torque rise time, 5 s. (b) Torque rise time, 1 s.

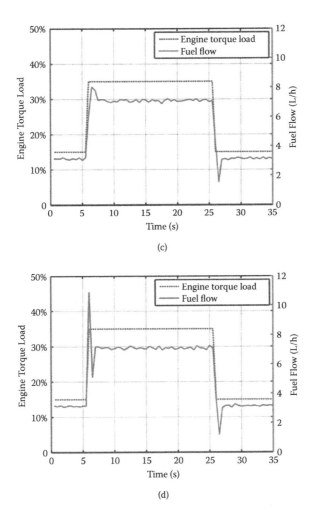

FIGURE 4.9 (Continued)
Engine fuel injections under different torque transients. (c) Torque rise time, 0.5 s. (d) Torque rise time, 0.2 s. (From Y. Wang et al., *Journal of Automobile Engineering*, 227(11): 1546–1561, 2013.)

relationship and maintain a relatively simple structure, the Hammerstein model structure [20, 21], which contains a static nonlinearity in series with a linear dynamic system, is utilized for emission modeling. A general Hammerstein model can be described in the discrete-time form, where the static nonlinearity is approximated as a finite polynomial expansion:

$$y(K) + m_1 y(K-1) + m_2 y(K-2) + \ldots + m_N y(K-N)$$

$$= n_1 x(K-1) + n_2 x(K-2) + \ldots + n_N x(K-N) \tag{4.10}$$

$$x(K) = \gamma_1 u(K) + \gamma_2 u^2(K) + \gamma_3 u^3(K) + \ldots + \gamma_M u^M(K)$$

where u, x, and y are the system inputs, states, and outputs, respectively, $M, N \in N$, K is the discrete-time step, $m_1 \ldots m_N$ and $n_1 \ldots n_N$ and are constant.

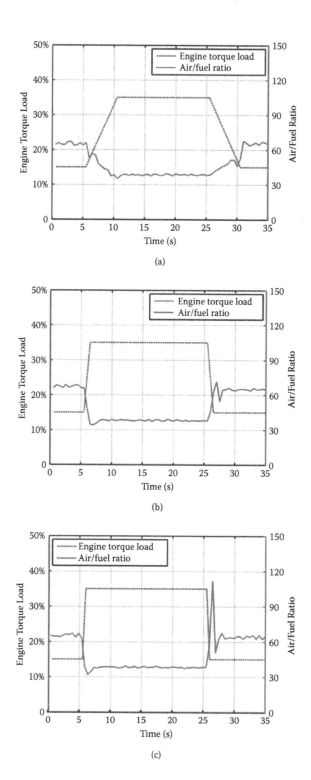

FIGURE 4.10
Engine air/fuel ratios under different torque transients. (a) Torque rise time, 5 s. (b) Torque rise time, 1 s. (c) Torque rise time, 0.5 s.

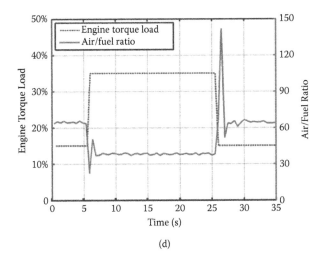

(d)

FIGURE 4.10 (*Continued*)
Engine air/fuel ratios under different torque transients. (d) Torque rise time, 0.2 s. (From Y. Wang et al., *Journal of Automobile Engineering*, 227(11): 1546–1561, 2013.)

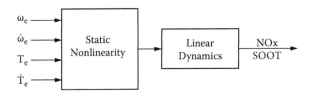

FIGURE 4.11
Hammerstein model structure. (From Y. Wang et al., *Journal of Automobile Engineering*, 227(11): 1546–1561, 2013.)

The block diagram of the proposed control-oriented emission model is shown in Figure 4.11 [25]. The static nonlinearity is presented by a third-order polynomial that consists of four input variables, which are proved to be sufficient to describe the relatively smooth ramp and saddle surfaces of the steady-state emissions shown in Figure 4.8(b, c). Besides, the linear dynamics is all set to be of first order, based on the autoregressive analysis of experimental data shown in the next section. Although there are multiple gaseous and particulate emissions, here we focus on the micro soot emission, and other emissions can be treated in a similar fashion.

The control-oriented soot emission model is shown as

$$\dot{Soot} = -\frac{1}{\tau} Soot + \frac{1}{\tau} \left[\lambda_1 T_e + \lambda_2 \omega_e T_e + \lambda_3 \dot{\omega}_e T_e + \lambda_4 \omega_e^2 T_e + \lambda_5 \dot{\omega}_e^2 T_e + \cdots + \lambda_{34} \omega_e \dot{T}_e^2 \right] \quad (4.11)$$

where the time constant τ and parameters $\lambda_1, \ldots, \lambda_{34}$ are all constants. If necessary, different time constants can be assigned to different terms of the input.

The emission model is identified using experimental data with the autoregressive (AR) algorithm. All the parameters $\tau, \lambda_1, \ldots, \lambda_{34}$ in the emission model described in Equation (4.11) are identified using experimental measurement of the input variables and soot emission. Figure 4.12 [25] shows the comparison of the micro soot emissions measured in experiments and predicted by the emission model,

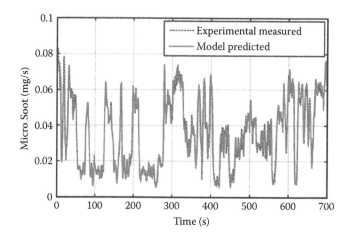

FIGURE 4.12
Experimental and model predicted emissions. (From Y. Wang et al., *Journal of Automobile Engineering*, 227(11): 1546–1561, 2013.)

along a predefined engine operation trajectory. It is obvious that the predictions based on the dynamic model match the experimental data very well.

In order to apply the proposed real-time optimal control to reduce local emissions, the emission model in Equation (4.11) needs to be further simplified. As will be shown later, the engine torque T_e and $\dot{T}_e$ torque rate are treated as state and control variables, respectively. Thus, the polynomials that contain the crossing terms of T_e and $\dot{T}_e$, and the polynomials that only contain the speed ω_e or acceleration $\dot{\omega}_e$ but without T_e or $\dot{T}_e$, should be avoided. Consequently, we simplify the soot emission model into the form given by

$$
\begin{aligned}
\dot{Soot} = & -\frac{1}{\tau}Soot \\
& + \frac{1}{\tau}\left[\lambda_1 T_e + \lambda_2 \omega_e T_e + \lambda_3 \dot{\omega}_e T_e + \lambda_4 \omega_e^2 T_e + \lambda_5 \dot{\omega}_e^2 T_e + \lambda_6 \omega_e \dot{\omega}_e T_e\right] \\
& + \frac{1}{\tau}\left[\lambda_7 \dot{T}_e + \lambda_8 \omega_e \dot{T}_e + \lambda_9 \dot{\omega}_e \dot{T}_e + \lambda_{10}\omega_e^2 \dot{T}_e + \lambda_{11}\dot{\omega}_e^2 \dot{T}_e + \lambda_{12}\omega_e\dot{\omega}_e\dot{T}_e\right]
\end{aligned}
\tag{4.12}
$$

where the parameters τ, $\lambda_1,\ldots,\lambda_{12}$ are constant.

Compared with the original model, the performance of the simplified emission model is degraded to some extent, but still acceptable, as shown in Figures 4.13 and 4.14 [25].

4.3.2.1.2 Two-Mode Hybrid Energy Management Strategy

To reduce the transient engine emissions without losing the global fuel efficiency optimization and battery SOC sustainability, a two-mode hybrid energy management strategy is presented, as shown in Figure 4.15 [25].

Along the predefined driving cycle, most of the time the hybrid vehicle is running in the fuel efficiency-improving mode, where both the engine speed and torque will be

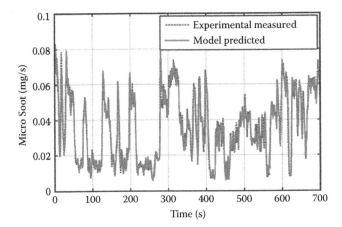

FIGURE 4.13
Experimental and simplified model predicted emissions. (From Y. Wang et al., *Journal of Automobile Engineering*, 227(11): 1546–1561, 2013.)

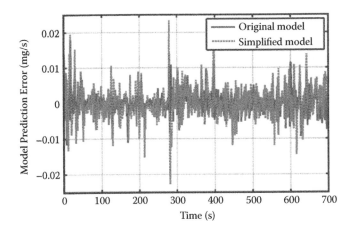

FIGURE 4.14
Original and simplified model prediction errors. (From Y. Wang et al., *Journal of Automobile Engineering*, 227(11): 1546–1561, 2013.)

optimized by the DP optimal control algorithm [4–9]. The cost of the DP optimization is calculated based on the steady-state fuel efficiency map (or the weighted fuel efficiency and emission map, if needed), so as to ensure the global optimization of the steady-state fuel consumption. However, since the DP optimization does not take the transient emissions into consideration, the optimized control may sacrifice the emission performance to a large extent, especially when the engine torque changes abruptly. To reduce the transient emission concentration, when the control commands from the DP optimization exceed some predefined threshold (i.e., the engine torque is forced to change abruptly), the emission-reducing mode is triggered and a local linear quadratic regulator (LQR) optimal control will override the global DP optimization by modifying the engine torque commands (i.e., fuel injection commands), but still holding the speed commands calculated by the DP. Generally, the LQR control will smooth the surging engine torque to reduce the high local emission concentration. This approach fully takes advantage of the hybrid

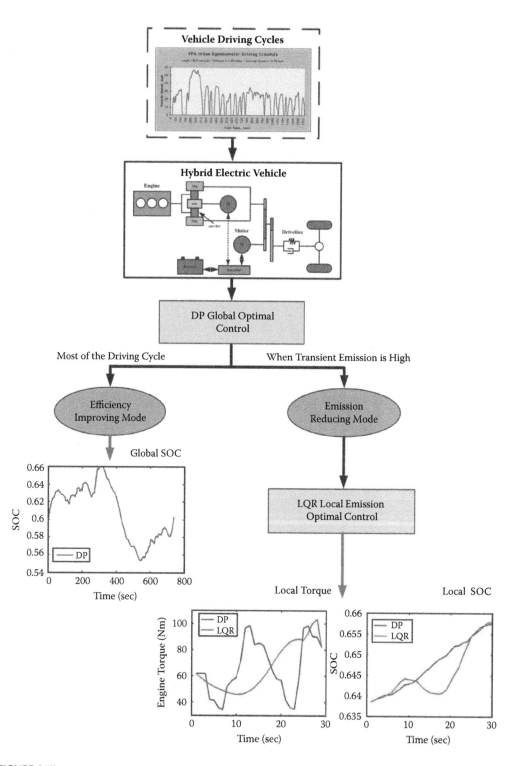

FIGURE 4.15
Two-mode fuel efficiency and emissions optimal control. (From Y. Wang et al., *Journal of Automobile Engineering*, 227(11): 1546–1561, 2013.)

powertrain system where the electric torques can be flexibly controlled to compensate the deficit engine torque at any time instant so as to keep tracking the predefined engine speed trajectory. At the end of the time horizon, the battery SOC of the hybrid vehicle will be driven back to the globally optimized point, so that the global fuel efficiency optimization and battery SOC sustainability can still be maintained. The design of the two-mode hybrid energy management strategy is shown in detail as follows:

1. Hybrid powertrain system dynamics. A typical power-split hybrid powertrain system (THS II), where a planetary gear set is employed as the core mechanism, is shown in Figure 4.16 [25]. The dynamic models for the power-split hybrid system have been shown in Section 4.3.1.

   With respect to the optimal target, the powertrain dynamics in Equations (4.1) to (4.8) can be written by a compact form:

$$\dot{\omega}_v = \frac{a_2}{a_1} T_e + \frac{a_3}{a_1} T_g + \frac{a_4}{a_1} T_m - \frac{a_4}{a_1}\left(\alpha\omega_v^2 + \beta\right)$$

$$\dot{\omega}_e = \frac{b_2}{b_1} T_e + \frac{b_3}{b_1} T_g + \frac{b_4}{b_1} T_m - \frac{b_4}{b_1}\left(\alpha\omega_v^2 + \beta\right) \tag{4.13}$$

$$\dot{SOC} = -\frac{V_{batt}}{2Q_{batt}R_{batt}} + \frac{\sqrt{V_{batt}^2 - 4\left(-T_g\omega_g\eta_g^{k_1} + T_m\omega_m\eta_m^{k_2}\right)R_{batt}}}{2Q_{batt}R_{batt}}$$

where

$$\omega_m = \omega_v K_{ratio}$$

$$\omega_g S + \omega_m R = \omega_e (R + S)$$

   The parameters a_1 to a_4, b_1 to b_4 are explained in [15] in detail.

2. Efficiency-improving mode: global DP optimization. In the efficiency-improving mode, dynamic programming, which has been widely investigated in the hybrid

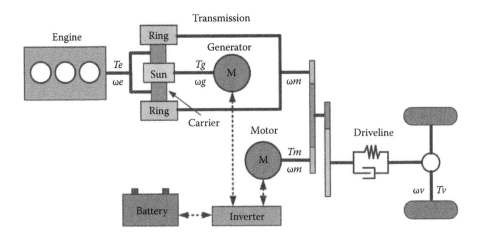

FIGURE 4.16
Power-split hybrid powertrain system. (From Y. Wang et al., *Journal of Automobile Engineering*, 227(11): 1546–1561, 2013.)

powertrain control area [4–9], is utilized as the energy management strategy. DP is a global optimization algorithm that backward searches all the possible states and feasible control actions along a predefined path, so as to find the optimal operation trajectory to meet the specific optimization objective (cost function).

The cost function used in the DP optimization is designed to minimize the fuel consumption and keep the SOC sustained in a limited bound, given by

$$
J = \sum_{k=0}^{N-1} \left\{ FC(k) \right\} + f_{soc} \left[SOC(N) - SOC(N)_{des} \right]^2
$$

$$
- m_{soc} \left[SOC(N) - SOC(N)_{des} \right]
$$

(4.14)

where FC is the fuel consumption in a single sampling step, $SOC(N)$ and $SOC(N)_{des}$ are the actual and desired battery SOC at the final step, f_{soc} is the SOC sustaining factor, and m_{soc} is the SOC variation compensation factor. Usually, f_{soc} will be given by a large value to force the SOC to converge to the desired value at the end; m_{soc} will be estimated based on the actual equivalent fuel consumption from the SOC deviation.

3. Emission-reducing mode: local LQR optimal control. In the emission-reducing mode, a local optimal control algorithm is required to realize two tasks:

 a. It modifies the engine torque trajectory to realize the local emission optimization, without changing the engine speed during the local time horizon.

 b. At the end of the local time horizon, the battery SOC should be driven back to the trajectory defined by the DP, so that the SOC sustainability provided by the DP can still be maintained.

To meet the above requirements, a linear quadratic regulator is utilized in the emission-reducing mode. LQR is a linear state feedback optimal control along some finite time period [22]. Thus, the linearization and necessary simplification of the nonlinear plant model, which includes the powertrain dynamics in Equation (4.13) and emission dynamics in Equation (4.12), are necessary.

For the engine and vehicle dynamics, the engine speed and vehicle speed will be maintained the same as the DP-optimized results; in other words, both ω_e and ω_v, as well as their time derivative, can be treated as known terms in the LQR control. Then the nonlinear terms involved with ω_e and ω_v in Equation (4.13) are actually eliminated, and we can rewrite the engine and vehicle dynamics as

$$
T_m = -\frac{a_2 b_3 - b_2 a_3}{a_4 b_3 - b_4 a_3} T_e + \frac{a_1 b_3 \dot{\omega}_v - b_1 a_3 \dot{\omega}_e}{a_4 b_3 - b_4 a_3} + \left(\alpha \omega_v^2 + \beta \right)
$$

$$
= r_{m1} T_e + r_{m2}
$$

$$
T_g = -\frac{a_2 b_4 - b_2 a_4}{a_3 b_4 - b_3 a_4} T_e + \frac{a_1 b_4 \dot{\omega}_v - b_1 a_4 \dot{\omega}_e}{a_3 b_4 - b_3 a_4}
$$

$$
= r_{g1} T_e + r_{g2}
$$

(4.15)

For the battery dynamics, if we neglect the internal resistance of the battery, the nonlinear dynamics can be transformed into a linear form:

$$
\dot{SOC} = -\frac{P_{batt}}{V_{batt}Q_{batt}} = -\frac{-T_g\omega_g\eta_g^{k1} + T_m\omega_m\eta_m^{k2}}{V_{batt}Q_{batt}}
$$

$$
= \frac{\left(r_{g1}T_e + r_{g2}\right)\left(\dfrac{\omega_e\left(R+S\right) - \omega_v K_{ratio}R}{S}\right)\eta_g^{k1}}{V_{batt}Q_{batt}}
$$

$$
\quad - \frac{\left(r_{m1}T_e + r_{m2}\right)\left(\omega_v K_{ratio}\right)\eta_m^{k2}}{V_{batt}Q_{batt}}
\tag{4.16}
$$

For the micro soot emission model shown in Equation (4.12), since the engine speed and its time derivative are both considered known parameters, the emission dynamics can be expressed by a more compact equation:

$$
\dot{Soot} = -\frac{1}{\tau}Soot + \frac{1}{\tau}\left[f_1\left(\omega_e, \dot{\omega}_e\right) + \cdots + f_6\left(\omega_e, \dot{\omega}_e\right)\right]T_e
$$

$$
\quad + \frac{1}{\tau}\left[f_7\left(\omega_e, \dot{\omega}_e\right) + \cdots + f_{12}\left(\omega_e, \dot{\omega}_e\right)\right]\dot{T}_e
\tag{4.17}
$$

where the functions $f_1, \ldots, f_6$ are λ_1, $\lambda_2\omega_e$, $\lambda_3\dot{\omega}_e$, $\lambda_4\omega_e^2$, $\lambda_5\dot{\omega}_e^2$, and $\lambda_6\omega_e\dot{\omega}_e$, respectively, and the functions $f_7, \ldots, f_{12}$ are λ_7, $\lambda_8\omega_e$, $\lambda_9\dot{\omega}_e$, $\lambda_{10}\omega_e^2$, $\lambda_{11}\dot{\omega}_e^2$, and $\lambda_{12}\omega_e\dot{\omega}_e$, respectively.

Further, combining Equations (4.15) to (4.17) yields a complete description of the engine soot emission dynamics (here a discrete-time form is used for the discrete-time LQR control):

$$
x(K + 1) = A(K)\, x(K) + B(K)\, u(K) + G(K)
$$

or

$$
x(K+1) = \begin{bmatrix} 1 & 0 & 0 \\ A_{21}(K) & A_{22}(K) & 0 \\ A_{31}(K) & 0 & 1 \end{bmatrix} x(K)
$$

$$
\quad + \begin{bmatrix} 1 \\ B_2(K) \\ 0 \end{bmatrix} u(K) + \begin{bmatrix} 0 \\ 0 \\ G_3(K) \end{bmatrix}
\tag{4.18}
$$

where the $x(K) = [T_e(K) \quad Soot(K) \quad SOC(K)]^T$ states and the input $u(K) = dT_e(K)$, which is the torque change from the K_{th} step to the $(K + 1)_{th}$ step. Other parameters are defined as

$$A_{21} = \frac{1}{\tau}\left[f_1\left(\omega_e, \dot{\omega}_e\right) + \cdots + f_6\left(\omega_e, \dot{\omega}_e\right)\right]$$

$$A_{22} = 1 - \frac{1}{\tau}$$

$$A_{31} = \frac{\left(r_{g1}\right)\left(\dfrac{\omega_e(K)(R+S) - \omega_v(K)K_{ratio}R}{S}\right)\eta_g^{k_1}}{V_{batt}Q_{batt}}$$

$$- \frac{\left(r_{m1}\right)\left(\omega_v(K)K_{ratio}\right)\eta_m^{k_2}}{V_{batt}Q_{batt}} \tag{4.19}$$

$$B_2 = \frac{1}{\tau}\left[f_7\left(\omega_e, \dot{\omega}_e\right) + \cdots + f_{12}\left(\omega_e, \dot{\omega}_e\right)\right]$$

$$G_3 = \frac{\left(r_{g2}\right)\left(\dfrac{\omega_e(K)(R+S) - \omega_v(K)K_{ratio}R}{S}\right)\eta_g^{k_1}}{V_{batt}Q_{batt}}$$

$$- \frac{\left(r_{m2e}\right)\left(\omega_v(K)K_{ratio}\right)\eta_m^{k_2}}{V_{batt}Q_{batt}}$$

After transforming the original nonlinear plant model into the linear form in Equation (4.18), the LQR optimal control algorithm can be designed. The optimal objective of the LQR control is to minimize the cost function within a predefined time period (in the discrete-time form, correspondingly, $K = 0, \ldots, N$):

$$J = \frac{1}{2}x^T(N)Hx(N) + \frac{1}{2}\sum_{K=0}^{N-1}\left[x^T(K)Q(K)x(K) + u^T(K)Ru^T(K)\right] \tag{4.20}$$

where the gain matrices H, Q, and R can be customized to meet the specific optimal target.

With the objective of minimizing the soot emission by smoothing the torque change, as well as maintaining the battery SOC the same as the DP result at the final stage, the gain matrices H, Q, and R are designed as follows:

$$H = \begin{bmatrix} h_1 & & \\ & h_2 & \\ & & h_3 \end{bmatrix}, Q = \begin{bmatrix} 0 & & \\ & q_2 & \\ & & 0 \end{bmatrix}, R = r_1$$

These gain matrices need to be tuned to produce a good balance of the soot emission minimization and battery SOC sustainability. Usually, h_3 is much larger than h_1 and h_2 for driving the SOC convergence to the predefined value; q_2 needs to be tuned carefully to reduce the soot emission during the time horizon; and r_1 can be quite small (since the effects of the torque transients on the soot emission have been contained in the dynamics of state x_2 and penalized by matrix Q). For example, a group of typical parameters is given by $h_1 = 0.01$, $h_2 = 1$, $h_3 = 1e7$, $q_2 = 5000$, and $r_1 = 0.01$.

Then the LQR state feedback control law u^* can be generated as

$$u^*(N-K) = F(N-K)\left\{ x(N-K) + \begin{bmatrix} 0 \\ 0 \\ G_3(N-K) \end{bmatrix} \right\} \tag{4.21}$$

where

$$F(N-K) = -[R + B^T(N-K)\ P(K)\ B(N-K)]^{-1}$$

$$\times\ B^T(N-K)\ P(K)\ A(N-K)$$

$$P(K+1) = [A(N-K) + B(N-K)\ F(N-K)]^T\ P(K) \tag{4.22}$$

$$\times\ [A(N-K) + B(N-K)\ F(N-K)]$$

$$+\ F^T(N-K)\ RF(N-K)] + Q$$

The generated LQR optimal control law is not the same as the standard form because of the existence of the known time-varying term G. However, the special form of the matrices A, G, and Q (which satisfies $AG = G$ and $G^TQG = 0$) allows us to modify the control law without violating the derivation of the closed-loop form of the LQR control law.

4.3.2.1.3 Experimental Results and Analysis

To demonstrate the performance of the proposed two-mode hybrid energy management strategy, a large number of experimental studies have been conducted under a high-fidelity hybrid powertrain testing environment. In particular, a rapid prototyping hybrid powertrain research platform was employed [23, 24]. With the hardware-in-the-loop (HIL) engine testing architecture, this research platform utilized a high-bandwidth hydrostatic dynamometer to precisely mimic the dynamic behaviors of the tested hybrid powertrain system to interact with a real-world John Deere 4045HF Tier IV diesel engine, so as to create the high-fidelity engine operations as if it were under the actual HEV dynamic environment, as shown in Figure 4.17(a) [25]. On this basis, state-of-the-art transient fuel consumption and emission measurement instruments (including the AVL P402 Fuel Measurement System, AVL Micro Soot Sensing System, and AVL FTIR Emission Sampling and Measurement System) were employed for system output measurements, as shown in Figure 4.17(b) [25].

The basic principle of switching between the two operation modes is, since the fuel efficiency is not considered in the LQR control, both the local time horizon and

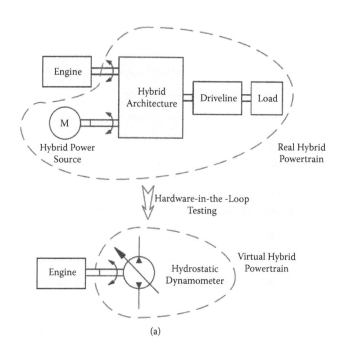

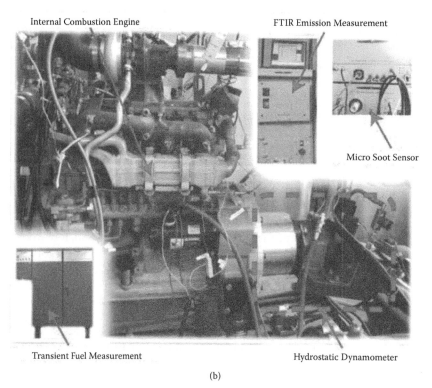

FIGURE 4.17
Hybrid powertrain research platform for HIL engine testing. (a) HIL system schematics. (b) System hardware and real-time measurement instruments. (From Y. Wang et al., *Journal of Automobile Engineering*, 227(11): 1546–1561, 2013.)

the accumulative time proportion of the emission-reducing mode in the whole driving cycle should be limited. In our experiments, when the DP-optimized engine torque changes rapidly above a threshold (>35 Nm/s), and the electric power usage is lower than its physical threshold (<20 kW), the emission-reducing mode is triggered; otherwise, the HEV will be operated in the fuel efficiency-improving mode. For each occurrence of the emission-reducing mode, the time duration will be in a range of 10 to 30 s.

Along a typical highway fuel efficiency testing (HWFET) cycle, the hardware-in-the-loop hybrid powertrain testing experiments with both the single-mode (DP only) and two-mode (DP plus LQR) energy management strategies are conducted. The experimental results, as shown in Figures 4.18 to 4.20 [25], demonstrate that the proposed two-mode energy management strategy can provide an obvious improvement on the soot emission compared with the single-mode strategy, without any significant loss in the fuel economy.

Figure 4.18(a) shows the number of occurrences of the emission-reducing mode along the HWFET driving cycle, which is determined by the switching rule and marked with the dash circle. In total, we get six emission-reducing events that will be named emission-reducing

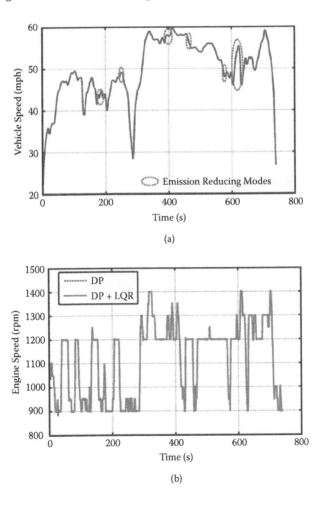

(a)

(b)

FIGURE 4.18
Desired vehicle speed and DP-optimized engine speed. (a) HWFET driving cycle. (b) Engine speed. (From Y. Wang et al., *Journal of Automobile Engineering*, 227(11): 1546–1561, 2013.)

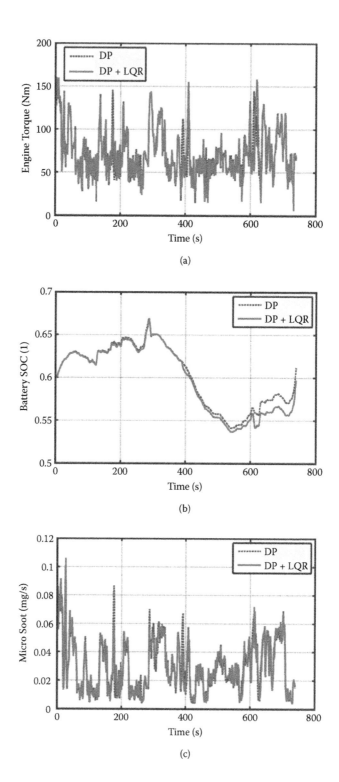

FIGURE 4.19
Globally and locally optimized trajectory along the HWFET cycle. (a) Engine torques. (b) Battery SOC. (c) Micro soot emission.

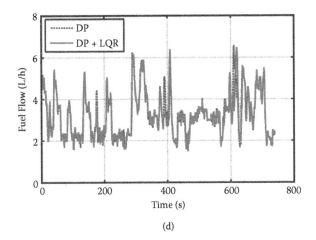

(d)

FIGURE 4.19 (*Continued*)
Globally and locally optimized trajectory along the HWFET cycle. (d) Fuel flow. (From Y. Wang et al., *Journal of Automobile Engineering*, 227(11): 1546–1561, 2013.)

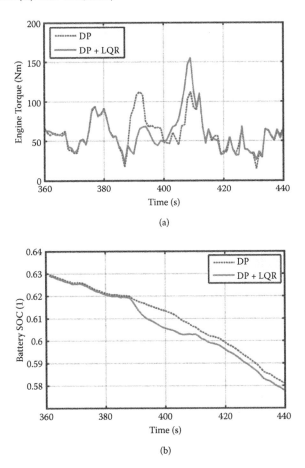

(a)

(b)

FIGURE 4.20
Zoomed-in globally and locally optimized trajectory between 360 and 440 s. (a) Engine torques. (b) Battery SOC.

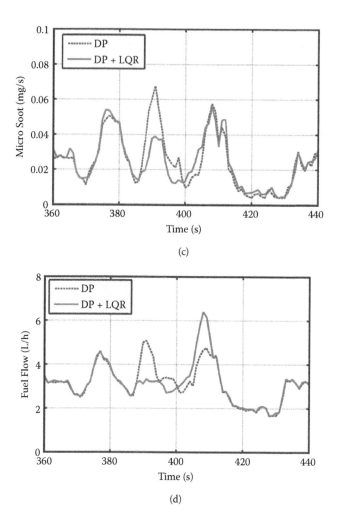

FIGURE 4.20 (*Continued*)
Zoomed-in globally and locally optimized trajectory between 360 and 440 s. (c) Micro soot emission.
(d) Fuel flow. (From Y. Wang et al., *Journal of Automobile Engineering*, 227(11): 1546–1561, 2013.)

event 1, event 2, ..., event 6. Figure 4.18(b) shows the DP-optimized engine speed trajectory
that will be tracked by both the single-mode and two-mode energy management strategies.

From Figure 4.19(a), the engine torque trajectory in the emission-reducing mode is quite
different from the one generated directly by the DP. However, due to the LQR optimal
control, the battery SOC will converge to the DP-optimized point at the final step, as
shown in Figure 4.19(b), so that the battery SOC is maintained in a sustainable fashion.
The difference in the engine torque trajectories leads to the difference in the micro soot
emissions and fuel consumption, as shown in Figure 4.19(c, d), which are further zoomed
in on in Figure 4.20. From Figure 4.20(c, d), in emission-reducing event 3 (387–410 s), the
soot emission is obviously reduced by the LQR, while the fuel consumption is maintained
at a similar level with the DP-optimized trajectories.

Tables 4.1 to 4.4 [25] further show the detailed data of the soot emissions, fuel consumptions, and battery SOC of all the emission-reducing events. From Table 4.1, the soot emission is reduced in every emission-reducing event and a total of 11.56% emission reduction is

TABLE 4.1

Emission Analysis of All the Emission-Reducing Events

Emission-Reducing Events	Micro Soot (DP)/mg	Micro Soot (LQR)/mg	Soot Reduction (by LQR)/%
1 (173–188 s)	0.4739	0.4581	3.33%
2 (256–266 s)	0.1150	0.0828	27.97%
3 (387–410 s)	0.8115	0.6742	16.92%
4 (462–472 s)	0.1249	0.1054	15.59%
5 (573–583 s)	0.2739	0.2239	18.24%
6 (606–636 s)	1.1715	1.0830	7.55%
Total	2.9707	2.6274	11.56%

TABLE 4.2

Fuel Consumption Analysis of All the Emission-Reducing Events

Emission-Reducing Events	Fuel Flow (DP)/ml	Fuel Flow (LQR)/ml	Fuel Flow Reduction (by LQR)/ml
1 (173–188 s)	11.7827	11.6668	0.1159
2 (256–266 s)	6.3150	5.8398	0.4752
3 (387–410 s)	24.7273	24.4847	0.2426
4 (462–472 s)	7.2152	6.7792	0.436
5 (573–583 s)	9.6822	8.9170	0.7652
6 (606–636 s)	34.9750	33.8021	1.1729
Total	94.6974	91.4896	3.2078

TABLE 4.3

Battery SOC Analysis of All the Emission-Reducing Events

Emission-Reducing Events	SOC Variation (DP)/1	SOC Variation (LQR)/1	SOC Loss (by LQR)/1
1 (173–188 s)	−0.0034	−0.0043	0.0009
2 (256–266 s)	0.0047	0.0037	0.0010
3 (387–410 s)	−0.0140	−0.0165	0.0025
4 (462–472 s)	−0.0020	−0.0028	0.0008
5 (573–583 s)	0.0041	0.0021	0.0019
6 (606–636 s)	0.0078	0.0002	0.0077
Total	−0.0028	−0.0176	0.0148

Note: Since the battery SOC varies between 0 (fully empty) and 1 (fully charged), its unit is set as 1 (i.e., normalized fully charged electric energy).

realized compared with the DP-only cases. Theoretically, the corresponding fuel consumption in the emission-reducing events can be slightly higher than the DP-only cases, since the LQR control does not include the fuel efficiency in the optimization cost function. However, the experimental fuel consumptions in the emission-reducing events are even lower than the DP-only case, as shown in Table 4.2. That can be explained by the deviations of battery SOC shown in Table 4.3. Because of the model uncertainty and disturbance during experiments, the battery SOC cannot be maintained at the absolutely same level

TABLE 4.4

Equivalent Fuel Consumptions of All the Emission-Reducing Events

Emission-Reducing Events	Transferred Fuel Flow from SOC Loss (by LQR)/ml	Equivalent Fuel Flow Increase (by LQR)/ml	Equivalent Fuel Flow Increase (by LQR)/%
1 (173–188 s)	0.2603	0.1444	1.23%
	~0.5422	~0.4263	~3.62%
2 (256–266 s)	0.2892	–0.186	–2.95%
	~0.6024	~0.1272	~2.01%
3 (387–410 s)	0.7229	0.4803	1.94%
	~1.5061	~1.2635	~5.11%
4 (462–472 s)	0.2313	–0.2047	–2.83%
	~0.4819	~0.0459	~0.64%
5 (573–583 s)	0.5494	–0.2158	–2.23%
	~1.1446	~0.3793	~3.92%
6 (606–636 s)	2.2266	1.0497	3.00%
	~4.6388	~3.4659	~9.90%
Total	4.2797	1.0719	1.13%
	~8.9161	~5.7083	~6.03%

as the DP-only cases. Eventually for the emission-reducing modes, the loss in the battery SOC will offset the gains in the fuel savings. In order to better evaluate the fuel consumptions with different management strategies, the SOC losses in Table 4.3 are transferred into the equivalent fuel consumptions by

$$FC_{equ} = \frac{\kappa_{trans} E_{elec} \Delta SOC \eta_{batt} \eta_{elec}}{Q_{LHV}} \qquad (4.23)$$

where E_{elec} is the full-charging electric energy of the storage battery, ΔSOC is the SOC loss, η_{batt} is the battery charging efficiency (here we use 90%), η_{elec} is the electric machine efficiency (here we use 85%), Q_{LHV} is the lower heating value of the diesel fuel (which is 43.4 MJ/kg), and κ_{trans} is the transfer factor from the mechanical energy to the fuel energy, i.e., the reciprocal of the conversion efficiency from the fuel energy (based on the lower heating value) to the real-world mechanical energy. It is not difficult to understand that the unpredictable distribution of the engine operations (torque/speed) in the engine map will introduce significant difficulty to accurately estimate the transfer factor. Therefore, instead of using a constant but inaccurate transfer factor, we use a range of transfer factors (corresponding to the largest/smallest engine conversion efficiencies in the engine operation area) to quantify the equivalent fuel consumption due to the SOC deviation. With an experimentally validated engine efficiency map, the transfer factor can be calculated within the range 2.9–6.0. Then Equation (4.23) will provide a range of equivalent fuel consumptions.

The equivalent fuel consumptions, as shown in Table 4.4, indicate that the LQR control introduces a little more fuel consumption than the DP-based control (within 1.13% to 6.03%), which is acceptable compared with the more significant emission reduction shown in Table 4.1. In conclusion, with the two-mode energy management strategy, the soot emission is significantly reduced without significant loss in the fuel economy.

4.3.2.2 DP-Based Extremum Seeking Energy Management Strategy

As an attempt to develop a fast and accurate hybrid energy management strategy, a stochastic dynamic programming–extremum seeking (SDP-ES) algorithm is designed [25]. This algorithm synthesizes the offline stochastic dynamic programming to ensure approximate global optimality and battery SOC sustainability, and employs real-time extremum seeking [26, 27] to provide a better local optimization. The SDP-ES algorithm combines the advantages of both SDP and ES, but loosens the limitations for each method, and hence reduces their disadvantages. This algorithm is characterized by two notable properties:

1. State-plus-output feedback: Not only the system states (engine speed, vehicle speed, and battery SOC) but also the system outputs (fuel consumptions and emissions) can be fed back to generate the control actions. In particular, the states are sent to the SDP control law, and the outputs are sent to the ES control law. This state-plus-output feedback can be realized with a rapid prototyping hybrid powertrain research platform, shown in Figure 4.17, where state-of-the-art fuel consumption and emission measurement instruments are employed.

2. Semi-model-based control: Although the input state models (hybrid powertrain dynamics and energy storage dynamics) can be attained to generate the SDP state feedback control law, it is difficult to obtain the accurate models between the outputs and states/inputs (engine combustion dynamics and efficiency maps). Here we apply the ES [26, 27], a non-model-based adaptive control, to online search the local optimal point by output feedback.

4.3.2.2.1 System Modeling

Again, the power-split HEV dynamic models are transformed into a compact form as described in Equation (4.13), which is further simplified as a third-order system:

$$\dot{x}_{HV} = f_{HV}\left(x_{HV}, u_{HV}\right) \tag{4.24}$$

where $x_{HV} = [\omega_v \quad \omega_e \quad SOC]^T$ are the states and $u_{HV} = [T_e \quad T_g \quad T_m]^T$ are the inputs. f_{HV} is the nonlinear system dynamics described in Equation (4.13).

Further, with the fuel consumption $\dot{m}_f$, micro soot emission $\dot{m}_{ms}$, and gaseous emission $\dot{m}_{gas}$ (or a single index that weights the three variables) as the system outputs, the dynamics between the outputs and inputs/states are given by

$$\dot{y}_{ICE} = g_{ICE}\left(y_{ICE}, x_{HV}, u_{HV}\right) \tag{4.25}$$

where $y_{ICE} = [\dot{m}_f \quad \dot{m}_{ms} \quad \dot{m}_{gas}]^T$ is the system outputs, and g_{ICE} denotes the dynamics between the system outputs and inputs (mainly T_e) plus states (mainly ω_e). In fact, the engine combustion dynamics in Equation (4.25) is asymptotically stable with respect to a group of equilibrium points. Hence, it is usually simplified to a static mapping $\bar{g}_{ICE}$ from an arbitrary input/state to the corresponding equilibrium point of the system outputs:

$$y_{ICE} = \bar{g}_{ICE}\left(x_{HV}, u_{HV}\right) \tag{4.26}$$

4.3.2.2.2 Analysis of the Real-Time Energy Management Strategies

4.3.2.2.2.1 Existing Real-Time Hybrid Energy Optimization Algorithms   Various hybrid energy optimization algorithms have been studied. The rule-based control [4, 5] employs a set of event-triggered rules to control the hybrid powertrain for high fuel efficiency, but it provides no guarantee of global optimality. The A-ECMS strategy [3, 10] converses the instantaneous electric power into the estimated equivalent fuel consumption to realize the instantaneous minimization of the gross fuel consumptions. A-ECMS simplifies the optimization problem; however, its optimality depends on the choice (or adaption) of the precise energy conversion factor, which is very difficult to attain for implementation, albeit existent in theory [28]. The MPC strategy [11, 12] formulates the global optimization problem into an optimization over a finite time window based on predictive models, and simplifies the nonlinear optimization to a linear quadratic program problem for which the analytical real-time solutions exist. The optimality of the MPC is limited by the assumption of the future vehicle speed (or power demand) in the prediction horizon.

Compared with the above algorithms, the SDP [6–9], which globally optimizes the fuel efficiency by searching all feasible control actions along an infinite time period (with a cost discounting factor) for all the possible states (with estimated probabilities) [6], presents its distinguished advantage in spite of huge offline computation. Theoretically, with the precise statistic knowledge of the future paths, the SDP is able to produce the optimal control in the statistic viewpoint [7]. Thus, it can be chosen as a method to approach the global energy optimal and battery SOC sustainable path.

4.3.2.2.2.2 Limitations of the SDP-Based Optimal Control   Even with a good expectation for the global energy optimality, there are some limitations that degrade the theoretical advantage of the SDP algorithm:

1. Curse of dimensionality: When the state variables increase in number, the offline computation load will increase exponentially. This greatly compounds the difficulty of the control implementation. Currently, the third-order power-split hybrid dynamic introduces three state variables and induces serious computation problems. To address this, some trade-offs are used to simplify the system models, e.g., limiting the engine power on a predefined curve. However, this will inevitably sacrifice the global optimality to some extent.

2. Negative effect of the large grid size: Similarly, with the purpose of reducing the computation load, the states and inputs are usually discretized with large grid size. This results in the generated optimal controls deviating from the actual optimal points due to the rough interpolations in implementation.

3. Lack of the precise system output model: Because of the lack of the complicated engine combustion dynamics, a static engine map that links the engine torque/speed with the fuel consumption is usually used for the offline SDP optimization.

4.3.2.2.3 Optimal Control Design

To make up the inherent deficiencies of the SDP algorithm, an output feedback-based optimization tool, extremum seeking (ES), is introduced to locally compensate the SDP optimal control, so as to drive the operating points toward the direction that further reduces fuel consumption. ES is a real-time optimization method that makes use of a periodic perturbation to dynamically search a maximum/minimum from an uncertain reference-to-output equilibrium map of a dynamic system. This dynamic system can be a nonlinear system if only its equilibrium map has a maximum/minimum, and all the equilibria are locally stable [26, 27].

4.3.2.2.3.1 SDP-ES Optimization Algorithm Design

The SDP-ES algorithm utilizes a SDP-based state feedback control as a reference term and injects a local feedback term via ES to compensate the control commands from SDP. The basic schematic of the SDP-ES optimization is shown in Figure 4.21 [25].

With the state feedback from the HEV system and speed (or power) demands from the driver model, the SDP will search through the control laws in an offline lookup table, to find a corresponding optimal engine power P_e, which further corresponds to an operating point $(\hat{\omega}_e, \hat{T}_e)$ on a predefined curve. Here, we call $(\hat{\omega}_e, \hat{T}_e)$ the SDP minimum, which means the minimal fuel consumption point calculated by the SDP. Based on a known SDP minimum, a SOC sustaining line can be generated based on the hybrid vehicle model [29, 30]. The battery SOC of all the points along this line can be maintained the same as the SDP minimum. From Equation (4.17), the SOC sustaining line is the collection of the points that satisfy

$$P_{elec} = -T_g\omega_g\eta_g^{k1} + T_m\omega_m\eta_m^{k2} = -\hat{T}_g\hat{\omega}_g\eta_g^{k1} + \hat{T}_m\hat{\omega}_m\eta_m^{k2} \tag{4.27}$$

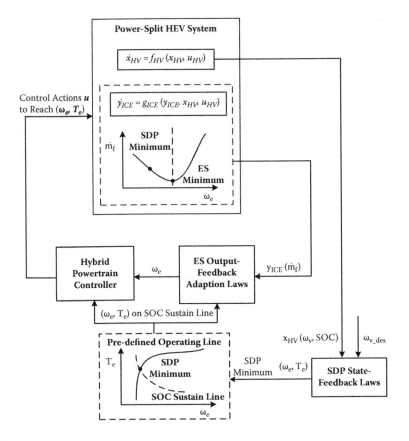

FIGURE 4.21
The schematic diagram of the SDP-ES optimization algorithm.

where $T_g\omega_g$ and $T_m\omega_m$ are the generator/motor powers of the arbitrary points on the SOC sustaining line, and $\hat{T}_g\hat{\omega}_g$ and $\hat{T}_m\hat{\omega}_m$ are the generator/motor powers of the SDP minimum.

To formulate the SOC sustaining line, the electric torques corresponding to the given vehicle speed ω_{v_des} and SDP minimum are given by

$$\hat{T}_g = \frac{\left[-\dfrac{a_1}{a_4}\dot{\omega}_{v\_des} + \dfrac{b_1}{b_4}\dot{\hat{\omega}}_e + \left(\dfrac{a_2}{a_4} - \dfrac{b_2}{b_4}\right)\hat{T}_e\right]}{\xi_1} \tag{4.28}$$

$$\hat{T}_m = \frac{\left[-\dfrac{a_1}{a_3}\dot{\omega}_{v\_des} + \dfrac{b_1}{b_3}\dot{\hat{\omega}}_e + \left(\dfrac{a_2}{a_3} - \dfrac{b_2}{b_3}\right)\hat{T}_e + \xi_2\left(\alpha\omega^2_{v\_des} + \beta\right)\right]}{\xi_2}$$

where

$$\xi_1 = b_3/b_4 - a_3/a_4, \quad \xi_2 = b_4/b_3 - a_4/a_3$$

Further, the electric power defined by the SDP minimum is

$$P_{elec} = -\hat{T}_g\left[\frac{R+S}{S}\hat{\omega}_e - \frac{R}{S}K_{ratio}\omega_{v\_des}\right]\eta_g^{k1} + \hat{T}_m K_{ratio}\omega_{v\_des}\eta_m^{k2} \tag{4.29}$$

Solving Equations (4.17), (4.27), and (4.29) yields the relationship between ω_e and T_e along the SOC sustaining line:

$$T_e = \frac{\left[\begin{array}{c} -P_{elec} - \sigma_1\xi_2\left(-\dfrac{a_1}{a_4}\dot{\omega}_{v\_des} + \dfrac{b_1}{b_4}\dot{\omega}_e\right) \\[2mm] +\sigma_2\xi_1\left[-\dfrac{a_1}{a_3}\dot{\omega}_{v\_des} + \dfrac{b_1}{b_3}\dot{\omega}_e + \xi_2\left(\alpha\omega^2_{v\_des} + \beta\right)\right] \end{array}\right]}{\sigma_1\xi_2\left(\dfrac{a_2}{a_4} - \dfrac{b_2}{b_4}\right) - \sigma_2\xi_1\left(\dfrac{a_2}{a_3} - \dfrac{b_2}{b_3}\right)} \tag{4.30}$$

where

$$\sigma_1 = \left[\omega_e(R+S)/S - K_{ratio}\omega_{v\_des}R/S\right]\eta_g^{k1}, \quad \sigma_2 = K_{ratio}\omega_{v\_des}\eta_m^{k2}$$

Equation (4.30) is a continuous function that formulates every engine operating point (ω_e, T_e) along the SOC sustaining line. Here, to ensure the SOC sustaining line is unique for a specific engine power, the engine torque T_e should be the only function of ω_e. To satisfy this condition, the term $\dot{\omega}_e$ in Equation (4.30) must be neglected [29, 30], which is equivalent to neglecting the engine motion dynamics. This simplification may induce some slight SOC deviations in the extremum seeking process from the SDP generated SOC. To compensate these deviations, some penalty terms about the transient power consumption for accelerating/decelerating the engine can be added on to the static fuel consumption, as part of the system output feedback.

Then, between two successive SDP commands, with the system outputs as the feedback, the ES control will adapt the optimal speed ω_e to minimize the fuel consumption (or emissions) along the calculated SOC sustaining line. Here, a hybrid powertrain controller [6, 31] is designed to generate the control u (engine throttle and torques T_m and T_g) to realize any engine operating point (ω_e, T_e) caught by the ES. After a short adaptive transient, the ES optimal control can find a new minimum point (ω_e^*, T_e^*), along the fuel consumption curve defined by SDP (i.e., the SOC sustaining line).

In summary, different from the state feedback and model-based SDP, the improved SDP-ES is a state-plus-output-feedback, semi-model-based optimization, which combines SDP and ES into one feedback controller. Essentially, the SDP can be treated as a reference component in the feedback control, which gives a control action profile based on the large amount of statistical data. This control profile can produce an approximate global optimal path from the statistic view and sustains the battery SOC. On this basis, without asking for any knowledge about the output dynamics $\bar{g}_{ICE}$ or the engine map, the ES compensates this SDP control profile using the instantaneous output measurement, to generate even better control actions. This is simply interpreted as follows.

In the discrete-time form, because the battery SOC generated by the SDP-ES is almost equal to the one from the SDP in every sampling step, the two accumulated SOC deviations at the final states will stay the same. Then, given a specific vehicle cycle, for the global cost function,

$$J = \sum_{k=0}^{N-1}\left\{FC(k)\right\} + f_{soc}\Delta SOC(N)^2 - m_{soc}\Delta SOC(N) \tag{4.31}$$

where FC is the fuel consumption in every sampling step, $\Delta SOC(N)$ is the error between the actual and the desired SOC at the final step, f_{SOC} is the SOC sustaining factor, and m_{SOC} is the SOC variation compensation factor. Only $FC(k)$ is concerned about the cost difference between SDP-ES and SDP.

Based on the optimal target of the ES control, for each step:

$$FC_{SDP-ES}(k) \leq FC_{SDP}(k), k \in [0, N-1] \tag{4.32}$$

From Equations (4.31) and (4.32), it is not difficult to conclude

$$\sum_{k=0}^{N-1}\left\{FC_{SDP-ES}(k)\right\} \leq \sum_{k=0}^{N-1}\left\{FC_{SDP}(k)\right\} \Rightarrow J_{SDP-ES} \leq J_{SDP} \tag{4.33}$$

Compared with the SDP strategy, the SDP-ES method will produce less fuel consumption along the whole driving cycle.

4.3.2.2.3.2 Design of the SDP Control Laws   The SDP algorithm is designed to generate the stationary optimal control policy π, which will directly map the driver's command (vehicle acceleration) and system states (SOC and vehicle speed ω_v) at the current time t to the control actions (engine power P_e) at time t. Based on various driving cycles, a Markov model is first designed to quantify the transition probability P_{ij} between different accelerations $\dot{\omega}_{v,j}$ and $\dot{\omega}_{v,i}$ with respect to a specific speed ω_v.

Although the plant model is a third-order system in Equation (4.17), the engine speed is coupled with the engine torque on a predefined operating line [6] for model simplification and the engine dynamics is eliminated. With the driver's command and current states (SOC, ω_v), a control P_e can be chosen and all the states at the next step can be solved with some transition probabilities. Then the cost function from state j to state i is

$$J_{\pi,i} = R\left(x_{HV}, u_{HV}\right)_i + \alpha \sum_j P_{ij} J_{\pi,j}$$

(4.34)

$$R\left(x_{HV}, u_{HV}\right)_i = FC\left(\omega_e, T_e\right)_i + f_{soc}\Delta SOC_i^2 - m_{soc}\Delta SOC_i$$

where $0 < \alpha < 1$ is the discount factor that determines the converge speed of the accumulated cost.

The SDP algorithm can be achieved through the policy iteration. Starting with an initial policy, the cost at state j will be calculated, and then for the policy improvement, a new policy at state i will be generated by

$$\pi_i = \arg\min\left[R\left(x_{HV}, u_{HV}\right)_i + \alpha \sum_j P_{ij} J_{\pi,j}\right]$$

(4.35)

Then, with the new policy, the cost will be updated and this algorithm will be performed iteratively, until π converges within an acceptable tolerance level due to the discount factor. Figure 4.22 [25] shows a map of a control law for a specific vehicle acceleration demand.

4.3.2.2.3.3 Design of the ES Output Feedback Control Laws Since any point (ω_e, T_e) on the SOC sustaining line can be positioned by one scalar parameter ω_e (or T_e), for the overall system that includes the hybrid powertrain and controller,

$$\dot{x}_{HV} = f_{HV}\left(x_{HV}, u_{HV}\right), \dot{y}_{ICE} = g_{ICE}\left(y_{ICE}, x_{HV}, u_{HV}\right)$$

we can use the control inputs u_{HV} (electric torques) to achieve the fast and precise track of desired state ω_e, and correspondingly produce a locally stable equilibrium y_{ICE}

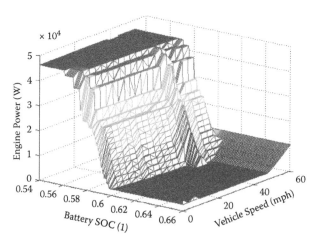

FIGURE 4.22
SDP optimal control law (1.08 mph/s acceleration demand).

(fuel consumption). On this basis, our optimization objective is to develop a feedback mechanism that adapts the optimal point ω_e, which minimizes the steady-state value of y_{ICE}, without asking for any knowledge of either f_{HV} or g_{ICE}. With this objective, a self-optimal ES control schematic is built, as shown in Figure 4.23 [25]. The basic design logic is shown below.

If we define the optimal speed corresponding to the minimal fuel consumption y_{ICE}^* as ω_e^*, then the estimate of ω_e^* can be defined as $\hat{\omega}_e$. For every $\hat{\omega}_e$ in the adaptive process, a periodic perturbation $a\sin(\omega t)$ is added onto the reference signal [26, 27]. This perturbation is designed slower than the plant dynamics (with the hybrid powertrain controller), so that the plant dynamics can be approximately treated as a static map $y_{ICE} = \phi(\omega_e)$, which will not seriously disturb the minimum seeking mechanism. When the simultaneous $\hat{\omega}_e$ is on the left side of the optimum ω_e^* ($\hat{\omega}_e < \omega_e^*$), the perturbation $a\sin(\omega t)$ will trigger a periodic response of $-y_{ICE}$ (here, $-$ is added to cope with the minimum seeking problem), which is in phase with $a\sin(\omega t)$; vice versa, when $\hat{\omega}_e$ is on the right side of the optimum ω_e^*, the triggered periodic output response will be out of phase with $a\sin(\omega t)$. Then, to avoid the DC component of $-y_{ICE}$ from interfering with the adaptive process, a high-pass filter $\dfrac{s}{s+\omega_h}$ is first designed to extract the high-frequency periodic components of $-y_{ICE}$. As a result, $\dfrac{s}{s+\omega_h}a\sin(\omega t)$ and $-\dfrac{s}{s+\omega_h}y_{ICE}$ will be approximately two sinusoidal signals that still satisfy: when $\hat{\omega}_e < \omega_e^*$, in phase; when $\hat{\omega}_e > \omega_e^*$, out of phase (we also add a high-pass filter for the perturbation $a\sin(\omega t)$ to compensate the phase deviation). Based on the trigonometric functions, the product of these two sinusoidal signals will have a DC component δ that satisfies: when $\hat{\omega}_e < \omega_e^*$, $\delta > 0$; when $\hat{\omega}_e > \omega_e^*$, $\delta < 0$. In particular, the DC δ will be approximately in the form of

$$\delta = \frac{-a^2\phi'\left(\hat{\omega}_e\right)}{2} \tag{4.36}$$

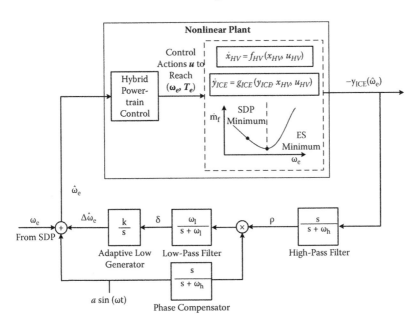

FIGURE 4.23
The control schematic of the ES real-time optimization.

After we extract δ by a low-pass filter $\dfrac{\omega_l}{s+\omega_l}$ [26, 27], the adaptive law to produce the compensation is given by $\Delta\hat{\omega}_e$

$$\Delta\dot{\hat{\omega}}_e = k\delta = \frac{-ka^2\phi'(\hat{\omega}_e)}{2} \tag{4.37}$$

Finally, the estimate $\hat{\omega}_e$ is generated by adding the ES compensation $\Delta\hat{\omega}_e$ and $a\sin(\omega t)$ onto a reference speed ω_e provided by the SDP. This adaptive control law can finally make the estimate $\hat{\omega}_e$ converge to the optimal point ω_e^*. Here, introducing a reference ω_e is to reduce the ES control efforts and make the adaptation converge quickly.

4.3.2.2.4 Simulation Results

To demonstrate the performance of the SDP-ES optimization algorithm, extensive simulation studies have been conducted along the highway fuel economy test (HWFET) driving cycle. An experimentally verified engine efficiency map of a 4.5 L John Deere diesel engine is employed to produce the system output. For the SDP control, the sampling rate is 1 Hz. For the ES control, a perturbation of $\sin(10\pi t)$ and the filters with $\omega_h = 100$ and $\omega_l = 2$ are used. An adaptive gain $k = 10,000$ is designed to speed up the seeking.

The simulation results demonstrate, for some driving scenarios where the SDP algorithm cannot produce the best results (e.g., an accelerating phase in the HWFET cycle, shown in Figure 4.24 [25]), the ES can provide a notable improvement on fuel efficiency compared with the original SDP. Here the offline global optimization results from the DP are also shown for reference. From Figure 4.24(a, b), it is obvious that both the SDP and SDP-ES can precisely track the vehicle speed and maintain the SOC at almost the same level (but different from the DP). Figure 4.24(c, d) shows the engine operating profiles (engine speed and torque) generated by the SDP and SDP-ES are quite different, which induces the difference on their fuel consumptions in Figure 4.24(e). In particular, SDP-ES improves the fuel efficiency at almost every time instant and finally achieves a 10% improvement based on the SDP (during this scenario, the fuel economy generated by SDP and SDP-ES are 45.72 and 50.14 mpg, respectively, and the theoretical optimal fuel economy from the DP is 53.63 mpg).

4.3.2.3 Driveline Dynamics Control for Hybrid Vehicles

The driveline dynamics of hybrid vehicles is more complex than that of conventional vehicles (see Chapter 3) due to the addition of the alternative power source and the more complex control. For example, the power-split hybrid system allows the engine power to be transmitted through both an electrical path and a mechanical path, and therefore improves the overall system efficiency. As a result of the optimized energy management, the internal combustion engine will start or stop more frequently than with conventional vehicles. During the engine start, however, significant torque pulsations will be generated due to the in-cylinder motoring/pumping pressure. The frequency of the engine torque pulsation is proportional to the engine speed, and the pulsation at low-speed range resonates with the driveline dynamics, and therefore leads to undesirable driveline vibration [32]. Similarly, the engine firing pulse, especially for advanced combustion with a short combustion duration, such as homogenous charge compression ignition (HCCI) [33, 34], will exhibit large torque pulsations and have the same issue for triggering driveline vibrations. To reduce the vibration, the energy management strategy needs to avoid certain operating conditions at the cost of sacrificing fuel economy. Therefore, it is desirable to remove or compensate the engine torque pulsations and broaden its range of operation to further improve vehicle fuel efficiency. The conventional approach is to add damping such as a torque converter, which results in energy loss. A promising approach is to control the electrical motor to track or reject the engine torque pulsation [32]. Due to the stroke-by-stroke motion of the internal combustion engine, the engine torque pulsation is naturally dependent on the rotational angle of the ICE. As shown in Figure 4.25 [35], the torque pulsation is periodic with respect to the rotational angle, but becomes aperiodic in the time domain as the engine speed varies. This unique feature suggests treating the problem in the angle domain, in which the period of the torque pulsation becomes invariant and the generating dynamics could be derived by leveraging the signal periodicity. It is possible to apply the internal model-based repetitive control framework to reject the pulsation. The merit of using an internal model controller is its ability to reject different kinds of engine torque oscillation, the exact shape of which is unknown in advance. However, due to the cost

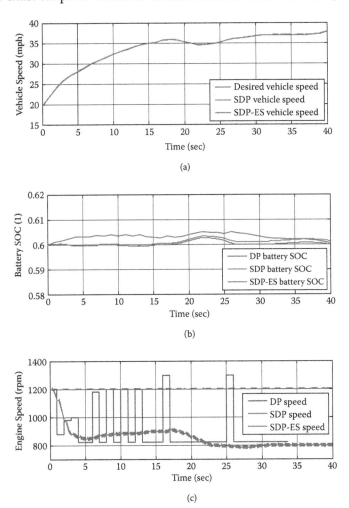

FIGURE 4.24
Comparisons between the optimized results by SDP and SDP-ES. (a) Vehicle speed. (b) Battery SOC. (c) Engine speed.

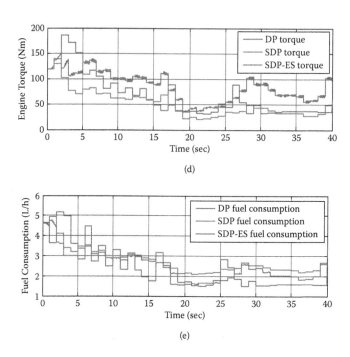

FIGURE 4.24 (*Continued*)
Comparisons between the optimized results by SDP and SDP-ES. (d) Engine torque. (e) Fuel consumption.

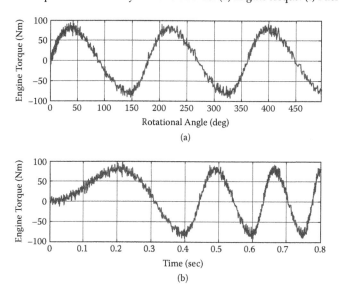

FIGURE 4.25
Engine torque oscillation in time and angle domain.

and accuracy concerns of the sensors, it is desirable to feed back the drive shaft speed oscillation instead of the torque vibration. The vibration of the drive shaft speed has the same frequency feature as the torque vibration, but its amplitude changes continuously. This unique feature again raises the necessity of looking into the tracking control for the periodic but amplitude-varying signal.

As revealed in [35], the generating dynamics of the magnitude-varying periodic signal could be time varying, instead of being time invariant, as those for a purely periodic signal. Thus, the traditional repetitive control design, which mainly treats time-invariant generating dynamics, cannot ensure asymptotic tracking performance. Therefore, a new framework for controlling magnitude-varying periodic signals needs to be used. More interestingly, in the hybrid powertrain problem, the magnitude variation for the velocity oscillation to be rejected is due to the integration of the torque vibration with varying frequency, and thus its generating dynamics can be derived. Furthermore, as shown in [35], when the hybrid powertrain vibration problem is treated in the angle domain, the actuator plant dynamics need to be converted to the angle domain as well, which will result in time-varying (actually angle-varying) plant dynamics. Therefore, the time-varying internal model-based control is a good fit for this application. A unique feature of the driveline vibration control for hybrid vehicles is the availability of the alternative power source, such as the motor/generator. By controlling the motor/generator, it is possible to suppress the driveline vibration and recover the vibration energy into the alternative power. Therefore, it is possible to apply active control to achieve desired driveline dynamics and improve vehicle fuel efficiency at the same time.

References

1. M. Ehsani, Y. Gao, and A. Emani, *Modern Electric, Hybrid Electric and Fuel Cell Vehicles, Fundamentals, Theory and Design*, 2nd ed., Boca Raton, FL: CRC Press, 2010.
2. A. Sciarretta and L. Guzzella, Control of Hybrid Electric Vehicles, *IEEE Control Systems Magazine*, 27(2): 60–70, 2007.
3. P. Pisu and G. Rizzoni, A Comparative Study of Supervisory Control Strategies for Hybrid Electric Vehicles, *IEEE Transactions on Control Systems Technology*, 15(3): 506–518, 2007.
4. C.C. Lin, H. Peng, J.W. Grizzle, and J. Kang, Power Management Strategy for a Parallel Hybrid Electric Truck, *IEEE Transactions on Control Systems Technology*, 11(6): 839–849, 2003.
5. H. Banvait, S. Anwar, and Y. Chen, "A Rule-Based Energy Management Strategy for Plug-In Hybrid Electric Vehicle (PHEV), in *Proceedings of the American Control Conference*, St. Louis, MO, 2009, pp. 3938–3943.
6. J. Liu and H. Peng, Modeling and Control of a Power-Split Hybrid Vehicle, *IEEE Transactions on Control Systems Technology*, 16(6): 1242–1251, 2008.
7. E. Tate, J. Grizzle, and H. Peng, SP-SDP for Fuel Consumption and Tailpipe Emissions Minimization in an EVT Hybrid, *IEEE Transactions on Control Systems Technology*, 18(3): 673–687, 2010.
8. S. Moura, H. Fathy, D. Callaway, and J. Stein, A Stochastic Optimal Control Approach for Power Management in Plug-In Hybrid Electric Vehicles, *IEEE Transactions on Control Systems Technology*, 19(3): 545–555, 2011.
9. J. Meyer, K.A. Stelson, A. Alleyne, and T. Deppen, Energy Management Strategy for a Hydraulic Hybrid Vehicle Using Stochastic Dynamic Programming, in *Proceedings of the 6th FPNI-PhD Symposium*, West Lafayette, IN, 2010, pp. 55–68.
10. C. Musardo, G. Rizzoni, Y. Guezennec, and B. Staccia, A-ECMS: An Adaptive Algorithm for Hybrid Electric Vehicle Energy Management, *European Journal of Control*, 11: 509–524, 2005.
11. H.A. Borhan, A. Vahidi, A.M. Phillips, M.L. Kuang, and I.V. Kolmanovsky, Predictive Energy Management of a Power-Split Hybrid Electric Vehicle, in *Proceedings of the American Control Conference*, St. Louis, MO, 2009, pp. 3970–3976.
12. T. Deppen, A. Alleyen, K.A. Stelson, and J. Meyer, A Model Predictive Control Approach for a Parallel Hydraulic Hybrid Powertrain, in *Proceedings of the American Control Conference*, San Francisco, CA, 2011, pp. 2713–2718.

13. C.D. Rakopoulos and E.G. Giakoumis, *Diesel Engine Transient Operation*, London: Springer, 2009.
14. J.R. Hagena, Z. Filipi, and D.N. Assanis, Transient Diesel Emissions: Analysis of Engine Operation during a Tip-In, SAE Technical Paper 2006-01-1151, 2006.
15. Y. Wang, H. Zhang, and Z. Sun, Optimal Control of the Transient Emissions and the Fuel Efficiency of a Diesel Hybrid Electric Vehicle, *Journal of Automobile Engineering*, 227(11): 1546–1561, 2013.
16. M. Benz, C.H. Onder, and L. Guzzella, Engine Emission Modeling Using a Mixed Physics and Regression Approach, *Journal of Engineering for Gas Turbines and Power*, 132(4): 042803, 2010.
17. Y. Wang, Y. He, and S. Rajagopalan, Design of Engine-Out Virtual NOx Sensor Using Neural Networks and Dynamic System Identification, *SAE International Journal of Engines*, 4(1): 837–849, 2011.
18. C. Ericson and B. Westerberg, Transient Emission Prediction with Quasi Stationary Models, SAE Technical Paper 2005-01-3852, 2005.
19. R. Ahlawat, J.R. Hagena, Z.S. Filipi, J.L. Stein, and H.K. Fathy, Volterra Series Estimation of Transient Soot Emissions from a Diesel Engine, in *IEEE Vehicle Power and Propulsion Conference (VPPC)*, Lille, France, 2010, pp. 1–7.
20. M. Adlouni, Modeling of Soot Emission for Heavy-Duty Diesel Engines in Transient Operation, Master's thesis, Lund University, Sweden, 2011.
21. E. Eskinat, S. Johnson, and W. Luyben, Use of Hammerstein Models in Identification of Nonlinear Systems, *AlChE Journal*, 37(2): 255–268, 1991.
22. D. Kirk, *Optimal Control Theory: An Introduction*, Upper Saddle River, NJ: Prentice Hall, 1970.
23. Y. Wang and Z. Sun, A Hydrostatic Dynamometer Based Hybrid Powertrain Research Platform, in *Proceedings of the International Symposium on Flexible Automation*, Tokyo, Japan, 2010, UPS-2739.
24. Y. Wang, Z. Sun, and K.A. Stelson, Modeling, Control and Experimental Validation of a Transient Hydrostatic Dynamometer, *IEEE Transactions on Control Systems Technology*, 19(6): 1578–1586, 2011.
25. Y. Wang and Z. Sun, SDP-Based Extremum Seeking Energy Management Strategy for a Power-Split Hybrid Electric Vehicle, in *Proceedings of the American Control Conference*, Montreal, Canada, 2012, pp. 553–558.
26. K. Ariyur and M. Krstic, *Real Time Optimization by Extremum Seeking Control*, New York: John Wiley & Sons, 2003, pp. 3–20.
27. M. Krstic and H. Wang, Stability of Extremum Seeking Feedback for General Nonlinear Dynamic Systems, *Automatica*, 36(2000): 595–601, 2000.
28. L. Serrao, S. Onori, and G. Rizzoni, ECMS as a Realization of Pontryagin's Minimum Principle for HEV Control, in *Proceedings of the American Control Conference*, St. Louis, MO, 2009, pp. 3964–3969.
29. P.Y. Li and F. Mensing, Optimization and Control of a Hydro-Mechanical Transmission Based Hybrid Hydraulic Passenger Vehicle, presented at Proceedings of the 7th International Fluid Power Conference, Aachen, Germany, 2010.
30. C.T. Li and H. Peng, Optimal Configuration Design for Hydraulic Split Hybrid Vehicles, in *Proceedings of the American Control Conference*, Baltimore, MD, 2010, pp. 5812–5817.
31. Y. Wang, X. Song, and Z. Sun, Hybrid Powertrain Control with a Rapid Prototyping Research Platform, in *Proceedings of the American Control Conference*, San Francisco, CA, 2011, pp. 997–1002.
32. S. Tomura, Y. Ito, K. Kamichi, and A. Yamanaka, Development of Vibration Reduction Motor Control for Series-Parallel Hybrid System, SAE Technical Paper 2006-01-1125, 2006.
33. T. Kuo, Valve and Fueling Strategy for Operating a Controlled Auto-Ignition Combustion Engine, in *SAE Homogeneous Charge Compression Ignition Symposium*, San Ramon, CA, 2006, pp. 11–24.
34. T. Kuo, Z. Sun, J. Eng, B. Brown, P. Najt, J. Kang, C. Chang, and M. Chang, Method of HCCI and SI Combustion Control for a Direct Injection Internal Combustion Engine, U.S. Patent 7,275,514, 2007.
35. X. Song, P. Gillella, and Z. Sun, Tracking Control of Periodic Signals with Varying Magnitude and Its Application to Hybrid Powertrain, in *Proceedings of the ASME Dynamic Systems and Control Conference*, Cambridge, MA, 2011, DSCC2010-4189.

5

Control System Integration and Implementation

5.1 Introduction to the Electronic Control Unit

An electronic control unit (ECU), also called the powertrain control module (PCM), is a type of electronic controller equipped with one or a few microprocessors that control a series of engine or transmission actuators based upon the sensor feedbacks to ensure the optimal engine or transmission performance. Before the electronic control became available, the engine air-to-fuel ratio, ignition timing, and idle speed were dynamically controlled by mechanical and pneumatic means.

5.1.1 Electronic Control Unit (ECU)

Modern ECUs use one or multiple microprocessor cores to process the inputs from the engine sensors and generate the corresponding control outputs for actuators in real time. An electronic control unit contains both the hardware and software. The hardware consists of electronic components on a printed circuit board (PCB), ceramic substrate, or thin laminate substrate. The main component on this circuit board is one or a few microcontrollers (CPUs). The software is often stored in the microcontroller(s) or other memory chips on the PCB, typically in so-called erasable programmable read-only memories (EPROMs) or flash memory so the CPU can be reprogrammed by uploading updated code or replacing chips. The engine control system is also called an (electronic) engine management system (EMS). Figure 5.1 shows a sample ECU architecture, with the sensors and actuators in the dotted boxes.

Besides managing the engine combustion process, a modern engine management system also controls many engine subsystems, such as variable valve timing subsystem, electronic throttle, exhaust gas recirculation (EGR) system, turbocharger and wastegate subsystems, and so on. The ECU also communicates with other control units, such as the transmission control unit and traction control unit, through the control area network (CAN).

5.1.1.1 ECU Control Features

There are many control features for an engine control system, including air-to-fuel ratio control, engine ignition timing control, EGR control, engine valve timing control, etc. The following is a list of these key control features.

1. Air-to-fuel ratio control. The engine air-to-fuel ratio is normally controlled by the injected fuel quantity, where engine load (or output torque) is normally controlled by the engine throttle (or charge air). Since the three-way catalyst provides

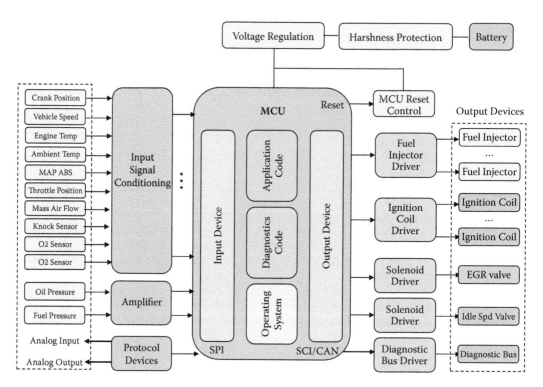

FIGURE 5.1
ECU architecture with sensors and actuators.

the highest conversion efficiency for hydrocarbon (HC), carbon monoxide (CO), and nitrogen oxides (NOx) at the stoichiometric air-to-fuel ratio (see Figure 5.2), in order to minimize the tailpipe emissions the gasoline engine needs to be operated at its air-to-fuel ratio close to the stoichiometric level, which is the main motivation for the air-to-fuel ratio regulation. The electronic control unit (ECU) determines the quantity of fuel to be injected based upon a number of parameters, such as acceleration pedal position, engine coolant temperature, engine speed, and so on. For the port fuel injection (PFI) fuel system, the transient air-to-fuel ratio control will be fairly challenging due to the so-called wall-wetting dynamics. Advanced control technology can be used to improve the accuracy of the air-to-fuel control during transient operations; see references [1–3].

2. Ignition (spark) timing control. A spark-ignited (SI) engine requires a spark to initiate the in-cylinder combustion process. An ECU adjusts the spark timing (or ignition timing) to optimize engine fuel economy with satisfactory emissions and improved power output. That is, if possible, the engine shall be operated at its maximum torque for the best torque (MBT) timing for the best fuel economy. As the result of engine downsizing, turbocharged engines are becoming popular and engine knock control becomes even more important. Besides engine EGR control, engine ignition timing is often used to control the engine knock; see references [4–6]. Since engine cold-start emissions consist of up to 80% of the total emissions of the Federal Test Procedure (FTP) driving cycle, reducing cold-start duration is very important for the SI engines. The spark timing is often retarded during the cold start to increase the exhaust temperature so that the three-way

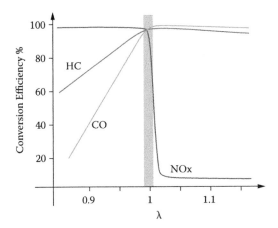

FIGURE 5.2
Conversion efficiencies of the three-way catalyst.

catalyst can be warmed up quickly. However, heavily retarded spark timing could lead to unstable combustion [4], and closed-loop control of the spark timing makes it possible to operate the engine at its maximum allowed retard timing with guaranteed combustion stability.

3. Idle speed control. Idle speed control is often part of the modern engine control system. The engine speed is measured by the so-called crankshaft position sensor, which also provides real-time engine position information for generating engine position-related control signals, such as ignition timing control, fuel injection control, and direct injection (DI) fuel pump control. Idle speed is controlled by an idle air bypass control stepper motor or a linear solenoid, while the engine throttle is positioned at its default limp-home position. The idle speed control must be robust to various engine load changes, such as air conditioning compressor or transmission engagement torque loads. Note that a full authority throttle control system can also be used to control engine idle speed and manage high-speed limitations.

4. Variable valve timing (VVT) control. To improve the engine fuel economy, most modern engines are equipped with VVT actuators to make it possible to adjust the engine intake or exhaust valve timings to their optimal positions for the best fuel economy. There are two kinds of VVT actuators: electrohydraulic and electric VVT. Both actuators regulate the relative position between the crank- and camshafts such that the intake or exhaust timing can be controlled. This feature can be used to optimize the charge airflow into the cylinder for the best fuel economy. The control of the electrohydraulic VVT [8] is through a proportional solenoid valve, and the control of the electric VVT [7] is by regulating the ring gear speed of the electric VVT planetary gear set through the electric motor. A cam position sensor is often used for closed-loop control.

5. Exhaust gas recirculation (EGR) control. The engine EGR is often used to reduce the engine NOx emissions, and recently it was also used to suppress the engine knock. Since a high EGR rate could lead to unstable combustion, to maximize the percentage of emission reduction, the EGR rate needs to be controlled in a closed loop to maintain the combustion stability [9]. The EGR rate is often controlled through an EGR valve driven by either a DC or step motor. A delta

pressure sensor is often used to estimate the EGR flow rate for closed-loop control. For the engine equipped with the turbocharger, the EGR can also be controlled by regulating the exhaust pressure through the variable geometry turbocharger or wastegate.

6. Variable geometry turbocharger and wastegate control. The turbocharger is used to boost intake manifold pressure to increase the engine power density and improve fuel economy. A variable geometry turbocharger is used to alter the turbo-map to best match the engine operational condition undercurrent speed and load condition. The wastegate is often used, along with the variable geometry turbocharger, to regulate the boost and exhaust pressures.

5.1.2 Communications between ECUs

Bosch originally developed the control area network (CAN) in 1985 for in-vehicle networks. Before that, automotive manufacturers connected electronic devices in vehicles using point-to-point wiring systems. As the number of electronic control devices increased, the wire harness got bulky, heavy, and expensive. It was then replaced by in-vehicle networks such as CAN to reduce wiring cost, complexity, and weight. CAN, a high-integrity serial bus system for networking intelligent devices, became the standard in-vehicle network. The automotive industry quickly adopted CAN, and in 1993, it became the international standard known as ISO 11898.

The main benefits of using CAN are the low cost and light weight. That is, the CAN network provides a low-cost, durable, and reliable network that enables communication among multiple CAN devices. The main advantage of CAN communication is that electronic control units (ECUs) can have a single CAN interface rather than using multiple wires to connect the analog and digital signals between these control units. This decreases overall cost and weight in automobiles. The following is a list of CAN communication features:

1. Communication broadcast. Each of the devices on the network has a CAN controller chip. All devices on the CAN network see all transmitted messages. Each device can decide if a message is relevant to itself. If not, it will be filtered. This structure allows modifications to CAN networks with minimal impact, and additional nontransmitting nodes can be added without modification to the network.

2. Priority. Each CAN message has a priority assigned such that if two nodes try to send messages simultaneously, the node with the higher priority has the right of transmission and the node with the lower priority gets postponed. This arbitration is nondestructive and results in noninterrupted transmission of the highest priority message. This also allows networks to meet deterministic timing constraints.

3. Error capability. The CAN specification includes a cyclic redundancy code (CRC) to perform error checking on the contents of each frame. Frames with errors are disregarded by all nodes, and an error frame can be transmitted to signal the error to the network. Global and local errors are differentiated by the controller, and if too many errors are detected, individual nodes can stop transmitting errors or disconnect themselves from the network completely.

4. CAN physical layers. CAN has several different physical layers you can use. These physical layers classify certain aspects of the CAN network, such as electrical

levels, signaling schemes, cable impedance, maximum baud rates, and more. The most common and widely used physical layers are described below:

- *High-speed CAN* is by far the most common physical layer. High-speed CAN networks are implemented with two wires and allow communication at transfer rates up to 1 Mbit/s. Other names for high-speed CAN include CAN C and ISO 11898-2. Typical high-speed CAN devices include antilock brake systems, engine control modules, transmission control systems, and emission systems.

- *Low-speed/fault-tolerant CAN networks* are also implemented with two wires, can communicate with devices at rates up to 125 Kbit/s, and offer transceivers with fault-tolerant capabilities. Other names for low-speed/fault-tolerant CAN include CAN B and ISO 11898-3. Typical low-speed/fault-tolerant devices in an automobile include comfort devices. Wires that have to pass through the door of a vehicle are low speed/fault tolerant in light of the stress that is inherent to opening and closing a door. Also, in situations where an advanced level of security is desired, such as with brake lights, low-speed/fault-tolerant CAN offers a solution.

- *Single-wire CAN interfaces* can communicate with devices at rates up to 33.3 Kbit/s (88.3 Kbit/s in high-speed mode). Other names for single-wire CAN include SAE-J2411, CAN A, and GMLAN. Typical single-wire devices within an automobile do not require high performance. Common applications include comfort devices such as seat and mirror adjusters.

- *Software-selectable CAN hardware*, such as National Instruments CAN hardware products, can be configured by software to select the desired CAN interface for the onboard transceivers among the high-speed, low-speed/fault-tolerant, and single-wire CANs. Multiple-transceiver hardware offers the perfect solution for applications that require a combination of communications standards.

CAN was first created for automotive applications. Therefore, the most common application is in-vehicle electronic networking. However, as other industries have realized the dependability and advantages of CAN communication over the past decades, it has been adopted in various applications. For example, the railway applications include different levels of the multiple networks, such as linking the door units or brake controllers, passenger counting units, and more. CAN also has applications in aircraft with flight state sensors, navigation systems, and so on. In addition, you can find CAN buses in many aerospace applications, ranging from in-flight data analysis to aircraft engine control systems such as fuel systems, pumps, and linear actuators.

The medical equipment industry also uses CAN as an embedded network in medical devices. Some hospitals use CAN to manage complete operating rooms, including control of the operating room components such as lights, tables, cameras, x-ray machines, and patient beds.

5.1.3 Calibration Methods for ECU

The engine and powertrain calibration process is always associated with high cost and long duration due to the system complexity. To determine the optimal powertrain and engine control parameters with respect to the desired performance, fuel economy, and emissions, the characteristics of the powertrain and engine system need to be accurately measured. Due to the high degrees of system freedom, a complete mapping is extremely time-consuming and often impossible. For instance, an engine with fuel injection, ignition timing, intake and exhaust VVT, EGR, and variable geometry turbocharger with wastegate can have up to seven degrees of freedom. Mapping the engine performance of seven control actuators with 10 test points per actuator requires 10^7 test points, which is impossible to be completed in a timely manner. Therefore, the use of model-based calibration methods [10] is a necessity. Design of experiments (DOE) is often used for the data-driven calibration process [11].

5.2 Control Software Development

Powertrain and engine control software is becoming increasingly sophisticated as automotive manufacturers strive to deliver improved fuel efficiency with reduced emissions. In particular, tighter emission regulations require precise control of the engine combustion process. With such a complex system and strict requirements, a model-based control development process is necessary for efficient control software development.

5.2.1 Control Software Development Process

The ECU hardware and software cannot be separated and the low-level software (see Figure 5.3) is the interface between the hardware and software. Normally the low-level driver software is provided by the ECU hardware provider, along with the real-time operating system (OS). The high-level driver either converts the sampled physical sensor signals into physical variables that can be used in the control features or converts the control variables into the signals such that the low-level driver can use them to interact with the ECU hardware to generate electronic control signals for the actuators. The control software development process is mainly associated with control feature development, shown in Figure 5.3 above the "Data Interface."

Traditionally, the control features were developed based upon the flowchart shown in Figure 5.4. It starts with control feature definition and concept development, followed by

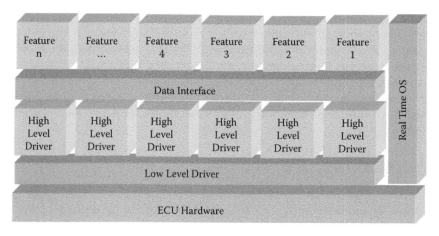

FIGURE 5.3
ECU hardware and software.

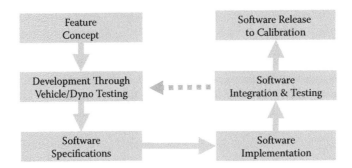

FIGURE 5.4
Traditional control feature development process.

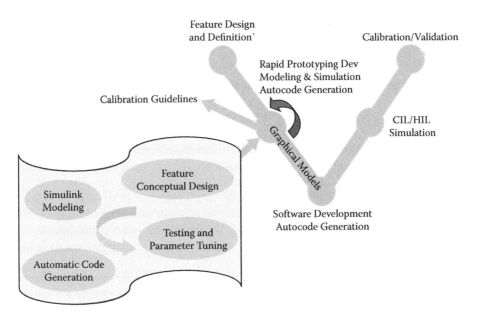

FIGURE 5.5
Rapid prototype (model-based) control development.

an iterative process of vehicle and dynamometer tests, software specification development and implementation, and software integration and testing. This process may iterate a few times until the software is ready for release. This is a fairly lengthy process since each iteration involves vehicle and dynamometer tests with manual software programming and implementation. Also, since it is a manual process, there is no mechanism to guarantee that the developed software will be consistent with the initial concept, and manual software debugging is also a concern. Therefore, the control feature development is moving toward a model-based process, as shown in Figure 5.5.

The model-based software development process is demonstrated in Figure 5.5, where the vehicle and dynamometer tests are replaced by simulations based upon the real-time (or control-oriented) engine model described in Chapter 2. During the control feature development process, the feature requirement will be documented by the graphical Simulink model, and the control feature is also developed and validated in a Simulink environment. Note that in order to validate the developed control feature, a control-oriented engine model (described in Chapter 2) is required. One of the features of this model-based development process is that the initial control feature calibrations can be generated during the feature validation process. Also, software autocode generation is part of the process that will be discussed in the next subsection.

Controller-in-the-loop (CIL) and hardware-in-the-loop (HIL) simulations are the key for validating the control features developed by the model-based process. The CIL simulations are conducted with a production ECU and an engine simulator, where the developed control feature is implemented into the production ECU. This is an important step since it validates the fact that the developed control feature can be executed in the production ECU, while the HIL simulation environment consists of the production ECU and a hybrid engine simulation platform that comprises a real-time simulator and part of the engine hardware, such as fuel injectors, ignition coils, etc. Therefore, CIL simulation is a special case of the HIL simulation, where no engine hardware is used in the simulation environment.

5.2.2 Automatic Code Generation

After the developed control feature has been validated to meet the specifications and requirements through software-in-the-loop (SIL) simulations (see Figure 5.5), the next step is to develop the software that can be deployed into the target ECU or an embedded processor. Since the control feature is developed in a MATLAB®/Simulink® environment, it is natural to use the MathWorks' Real-Time Workshop Embedded Coder (or Simulink Coder) to automatically generate highly efficient C/C++ code from the feature control model in Simulink. The code generated by the Embedded Coder is capable of running on virtually any microprocessor or real-time operating system (RTOS). The generated code can also be integrated back into Simulink for SIL simulation validation, and then deployed directly to a supported microprocessor for CIL and HIL simulation validation.

The main advantage of the autocode generation is its traceability among the Simulink-based feature requirements (specifications), the feature control model, feature test vectors used to validate the developed feature, and the C/C++ code generated by the Embedded Coder. This ensures that any design decision (selection) made earlier in the feature development process is fully documented within the generated code. The ability to navigate from the Simulink feature definition to the developed control feature in Simulink and to the embedded C/C++ code generated by Simulink Coder provides a systematic approach for control software development. It is common in the automotive industry that the control feature requirements need to be changed to meet future fuel economy and emission requirements. With the model-based development process, the modifications can be made in the Simulink-based control feature and the modified feature can be validated through the software-in-the-loop (SIL) simulations. After the feature is validated in SIL simulations, the code generation can be done automatically, which reduces the feature modification duration significantly, and hence reduces software development cost.

5.2.3 Software-in-the-Loop (SIL) Simulation

There are many definitions for SIL simulations. In this book, we call the co-simulation conducted on a single computer platform with both engine model and its controller the SIL simulation (see Figure 5.6). It can be divided into two main categories: model-in-the-loop (MIL) simulation and software-in-the-loop simulation. MIL simulations use a controller

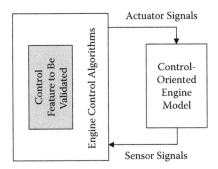

FIGURE 5.6
SIL simulation environment.

model (normally in Simulink) along with a plant (engine) model, and the control model is connected directly to a physical model of the system within the same Simulink diagram. In this case, any small modifications to the control model can be immediately tested in the MIL simulation environment, and it is also very easy to tune the control parameters with simulation validation so that the initial calibrations can be generated in this stage. The SIL simulation environment makes the simulation more realistic by replacing the Simulink-based control strategy with the strategy based upon the embedded C/C++ code. This type of simulation can be used to check if there is any difference between the Simulink-based and embedded code-based strategies by running the parallel simulations with both controllers at the same time. The SIL simulation is closer to the actual implementation since the target-embedded C/C++ code is executed in the simulations, which is essential for validating the coding system (whether autocoded or human coded).

5.2.4 Hardware-in-the-Loop (HIL) Simulation

As discussed before, the HIL simulation environment consists of the engine control model, engine model simulator, and engine system-associated hardware. This subsection concentrates on the controller-in-the-loop (CIL) simulations as shown in Figure 5.7, where an Opal-RT-based engine prototype controller was used to control the virtual engine simulated by a dSPACE simulator in a CIL simulation environment. In the rest of this subsection the charge air control feature of a homogeneous charge compression ignition (HCCI) engine is used to demonstrate the rapid prototype control development process discussed in this chapter.

5.2.4.1 HCCI Combustion Background

Homogeneous charge compression ignition (HCCI) combustion has the potential for internal combustion (IC) engines to meet the increasingly stringent emissions regulations with improved fuel economy [12]. The flameless nature of the HCCI combustion and its high dilution operation capability lead to low combustion temperature. As a result, the formation of NOx can be significantly reduced [13]. Furthermore, the HCCI engine is capable of unthrottled operation that greatly reduces pumping loss and improves fuel economy [14, 15].

On the other hand, the HCCI combustion has its own limitations. It is limited at high engine load due to the mechanical limit, and at low load due to misfire caused by the lack

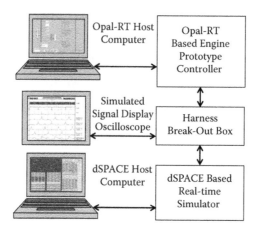

FIGURE 5.7
CIL simulation architecture.

of sufficient thermal energy to initiate the autoignition of the gas-fuel mixture during the compression stroke [16]. In fact, HCCI combustion can be regarded as a type of engine operating mode rather than a type of engine [17]. That is, the engine has to be operated in dual-combustion mode to cover the entire operational range.

It is fairly challenging to operate the engine in two distinct combustion modes, and it is even more difficult to have smooth combustion mode transition between SI and HCCI combustion because the favorable thermo conditions for one combustion mode are often adverse to the other [18]. For example, a high intake charge temperature is required in the HCCI mode to initiate the combustion, while in the SI mode it leads to reduced volumetric efficiency and increased knock tendency. For this reason, engine control parameters, such as intake and exhaust valve timings and lifts, throttle position, and EGR valve opening, are controlled differently between these two combustion modes. During the combustion mode transition, these engine parameters need to be adjusted rapidly. However, the physical actuator limitations on response time prevent them from completing their transitions within the required duration, specifically within one engine cycle. The multicylinder operation makes it challenging [19]. And this problem becomes more difficult when two-stage lift valve and electric VVT systems are adopted. Accordingly, the combustion performance during the transition cannot be maintained unless proper control strategy is applied.

The control of HCCI combustion has been widely studied in past decades. Robust HCCI combustions can be achieved through model-based control, as described in [20–22]. To make the HCCI combustion feasible in a practical SI engine, the challenge of the combustion mode transition is inevitable. In recent years, more and more attention has been paid to the mode transition control between SI and HCCI combustion. In [23] and [24], smooth mode transitions between SI and HCCI combustion are realized for a single-cylinder engine equipped with the camless variable valve actuation (VVA) system. However, high cost prevents the implementation of the camless VVA system in production engines. In [25] a VVT system with dual-stage lifts is used on a multicylinder engine for studying the mode transition. Experimental results show the potential of achieving smooth mode transition by controlling the step throttle opening timing and the direct injection (DI) fuel quantity. However, satisfactory mode transition has not been accomplished due to the lack of the robust mode transition control strategy.

The controller-in-the-loop (CIL) simulation results, shown below, demonstrated that unstable combustion during the transition can be eliminated by using the multistep strategy as discussed in [26] for a four-cylinder engine equipped with external cooled EGR, dual-stage valve lift, and electrical VVT system. The rest of this section utilizes the linear quadratic (LQ) optimal manifold air pressure (MAP) tracking control strategy to maintain the air-to-fuel ratio in the desired range so that hybrid (or spark-assisted) combustion is feasible. Under the optimal MAP control, smooth combustion mode transition can be achieved with the help of the iterative learning control (ILC) of the DI fuel quantity of individual cylinders. Note that the ILC is mainly used to generate transient fuel calibrations. The entire control strategy was validated in the CIL engine simulation environment [27], and satisfactory engine performance was achieved during the combustion mode transition for both steady-state and transient operating conditions.

5.2.4.2 Multistep Combustion Mode Transition Strategy

The configuration of the target HCCI-capable SI engine and the engine specifications are listed in Table 5.1. The key feature of this engine is its valvetrain system. It has two-stage lift for both intake and exhaust valves. The high lift has 9 mm maximum lift for the SI combustion mode, and the low lift has 5 mm maximum lift for the HCCI combustion mode. The ranges of both intake and exhaust valve timing are extended to ±40 crank degrees to improve the controllability of the internal EGR fraction, the effective compression ratio, and the engine volumetric efficiency during the combustion mode transition and HCCI operations. The externally cooled EGR is used to enable high dilution charge with a low charge mixture temperature.

For this application example, the combustion mode transition was studied for the engine operated at 2000 rpm with 4.5 bar indicated mean effective pressure (IMEP). Table 5.2 lists the engine parameters associated with the SI and HCCI combustion. These parameters were optimized for steady-state engine operation with the best fuel economy that satisfies the engine knock limit requirement. It can be seen in Table 5.2 that the optimized engine control parameters are quite different between the SI and HCCI combustion modes. Some of these parameters can be adjusted within one engine cycle, such as spark timing θ_{ST}, electronic throttle control (ETC) drive current I_{ETC} that is proportional to throttle motor torque, DI fuel quantity F_{DI}, and valve lift Π_{lift}; the others cannot due to slow actuator dynamics.

The combustion characteristics are also quite different between these two combustion modes, as illustrated in Figure 5.8. For example, the HCCI combustion has higher peak

TABLE 5.1

HCCI-Capable SI Engine Specifications

Engine Parameter	Model Value
Bore/stroke/con-rod length	86 mm/86 mm/143.6 mm
Compression ratio	9.8:1
Intake/exhaust valve lifts of high stage	9 mm/9 mm
Intake/exhaust valve lifts of low stage	5 mm/5 mm
Intake/exhaust valve timing range	±40°/±40°
Intake/exhaust valve lifts lash	0.2 mm/0.25 mm
Intake manifold volume	3.2 L
Throttle diameter	42 mm

TABLE 5.2

Engine Control Parameters for SI and HCCI Modes

Engine Control Parameter	SI	HCCI
θ_{ST} (° ACTDC)	−36	None
φ_{EGR} (%)	3	26
I_{ETC} (A)	0.84	5
F_{DI} (ms/cycle)	2.06	1.6
θ_{INTM} (° AGTDC)	70	95
θ_{EXTM} (° BGTDC)	100	132
Π_{lift} (mm)	9	5

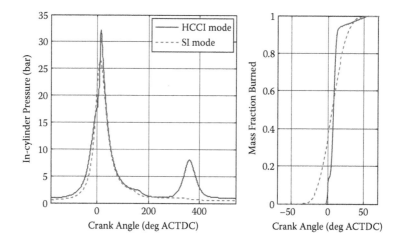

FIGURE 5.8
Steady-state combustion characteristics of SI and HCCI modes.

in-cylinder pressure than SI combustion due to the faster burn rate. Most likely, it also has a recompression phase (see the second peak of the solid line in Figure 5.8) due to the negative valve overlap (NVO) operation, while the SI combustion does not. The goal of the combustion mode transition is to switch the combustion mode without detectable engine torque fluctuations by regulating the engine control parameters, or in other words, to maintain the engine IMEP during the combustion mode transition.

The earlier work in [27] demonstrated that the engine charge temperature (T_{IVC}) has a response delay during the combustion mode transition, mainly caused by the response delays of the engine intake/exhaust valve timings and manifold filling dynamics. As a result, if the engine were forced to switch to the HCCI combustion mode, the engine IMEP could not be maintained with cycle-by-cycle fuel control F_{DI}. Also, the increased cooling effect caused by the increment of F_{DI} reduces the charge temperature and leads to unstable HCCI combustion. However, the transitional thermodynamic conditions are suitable for the SI-HCCI hybrid combustion mode proposed in [16] and [26].

By maintaining the engine spark (SI spark location), combustions during the mode transition could start in the SI combustion mode with a relatively low heat release rate, and once the thermo and chemical conditions of the unburned gas satisfy the start of HCCI (SOHCCI) combustion criteria, the combustion continues in HCCI combustion mode, which is illustrated by the solid curve of mass fraction burned (MFB), shown in Figure 5.9,

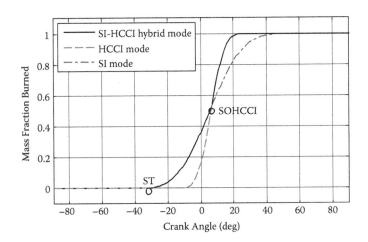

FIGURE 5.9
MFB trace of SI-HCCI hybrid combustion mode.

obtained through GT-Power simulations. During an ideal SI-to-HCCI combustion transition process, the HCCI combustion percentage (the vertical distance from SOHCCI to MFB = 1) increases gradually along with the gradual increase of charge temperature (T_{IVC}). For the HCCI-to-SI combustion transition, the HCCI combustion percentage will be gradually reduced. More importantly, during the SI-HCCI hybrid combustion, engine IMEP can be controlled by regulating the DI fuel quantity, which will be discussed later. This is the other motivation for utilizing the hybrid combustion mode during the combustion mode transition.

In [26], a crank-based SI-HCCI hybrid combustion model was developed for real-time control strategy development. It models the SI combustion phase under the two-zone assumption and the HCCI combustion phase under the one-zone assumption. The SI and HCCI combustion modes are actually special cases of the SI-HCCI hybrid combustion mode in the model, since the SI combustion occurs when the HCCI combustion does not occur, and the HCCI combustion occurs when the percentage is 100%. Accordingly, this combustion model is applicable for all combustion modes during the mode transition.

In [27], the one-step combustion mode transition was investigated. The control references of all engine parameters were directly switched from the SI mode to HCCI mode, as listed in Table 5.2, in one engine cycle. The simulation results showed that misfires occur during the one-step mode transition, and significant torque fluctuation was discovered. Thereby, a multistep mode transition strategy was proposed in [27] by inserting a few hybrid combustion cycles between the SI and HCCI combustion (see Figure 5.10). The proposed control strategy is based on this multistep strategy.

As illustrated in Figure 5.10, five engine cycles are used during the SI-to-HCCI mode transition. During the transitional cycles, some engine parameters are adjusted in open loop according to the schedule shown in Figure 5.10. Cycles 1 and 2 are used for engine throttle control. They provide enough time for the engine MAP to increase to compensate the valve lift (Π_{lift}) switch. At the end of cycle 2, the intake/exhaust valve lift Π_{lift} switches from high lift to low lift, and the control references of EGR fraction φ_{EGR}, intake valve timing θ_{INTM}, and exhaust valve timing θ_{EXTM} are set to those of the steady-state HCCI combustion mode as listed in Table 5.2. Spark timing θ_{ST} of each cylinder was kept constant

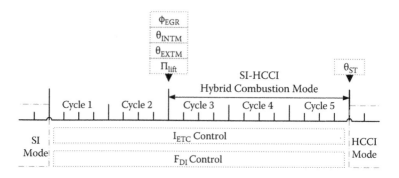

FIGURE 5.10
Multistep SI-to-HCCI combustion mode transition control schedule.

during the transitional cycles and was eliminated at the end of cycle 5. Throughout the transitional cycles, the engine control parameters, throttle current I_{ETC}, and DI fueling F_{DI} are regulated with the time-based control at a sample period of 1 ms and with cycle-based controls, respectively.

5.2.4.3 Air-to-Fuel Ratio Tracking Problem

To study the feasibility of using fuel injection quantity F_{DI} to regulate the engine IMEP, intensive simulations were conducted to map out the engine IMEP and air-to-fuel ratio as functions of engine fuel injection quantity F_{DI} and manifold air pressure (MAP). The simulation results are shown in Figure 5.11, indicating that the engine IMEP is highly correlated to F_{DI} with the lean air-to-fuel mixture. As a result, it is possible to control the individual cylinder IMEP by regulating the corresponding F_{DI}.

To maintain the controllability of the DI fueling (F_{DI}), a lean gas-fuel mixture is required during the mode transition. However, the combustion could become unstable if the mixture becomes extremely lean since the engine spark might not be able to ignite the gas mixture. For this study, the desired normalized air-to-fuel ratio is set between λ_{min} (0.97) and λ_{max} (1.3). In [27], a step throttle preopening approach was proposed to prevent rich combustions at cycle 3, but it leads to very lean combustion in the following engine cycles. For this application, an LQ tracking control strategy [28] is developed to regulate the air-to-fuel ratio around the desired level.

As discussed above, the normalized air-to-fuel ratio needs to be maintained within the optimal range ($\lambda_{min} \leq \lambda \leq \lambda_{max}$) during the SI-to-HCCI combustion mode transition. This control target is difficult to achieve through the air-to-fuel ratio feedback control due to delay and the short mode transition period. It is proposed to use the LQ optimal tracking approach to regulate the air-to-fuel mixture to the desired level. To implement this control strategy, the optimal operational range of λ is translated into the operational range of the engine MAP shown in Figure 5.12, where the upper limit corresponds to λ_{max} and the lower limit corresponds to λ_{min}. This provides an engine MAP tracking reference, shown in Figure 5.12, to maintain the engine MAP within the desired range. The reference signal is represented by

$$z(k) = \begin{cases} Z_{SI} & \text{if } k_B < k \leq k_1 \\ Z_{SI} + (Z - Z_{SI})\frac{k-k_1}{k_2-k_1} & \text{if } k_1 < k \leq k_2 \\ Z + (Z_{HCCI} - Z)\frac{k-k_1}{k_2-k_1} & \text{if } k_2 < k \leq k_E \end{cases} \tag{5.1}$$

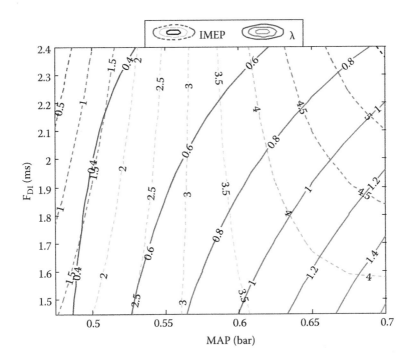

FIGURE 5.11
IMEP sensitivity analysis of the SI-HCCI hybrid combustion mode.

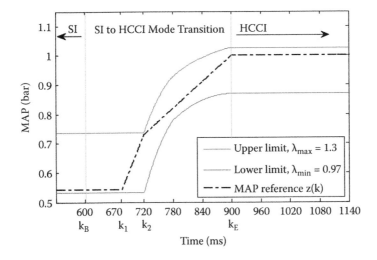

FIGURE 5.12
The target MAP operational range and MAP tracking reference.

where k is the time-based sampling index; k_B and k_E represent the beginning and ending indices of the mode transition, and they were set to 600 and 900, respectively, as shown in Figure 5.12; k_1 and k_2 are switch indices, and they equal 670 and 720, respectively; Z_{SI} and Z_{HCCI} are the desired MAPs of SI and HCCI modes, respectively; and Z is the desired MAP at k_2.

5.2.4.4 Engine Air Charge Dynamic Model

To develop the proposed LQ tracking control strategy, a simplified engine MAP model is required to represent the relationship between the control input (I_{ETC}) and the system output (MAP). The simplified dynamics are represented by the second-order dynamics due to the gas filling dynamics (first order) of the engine intake manifold and the first-order response delay of the engine throttle. The governing equation of gas filling dynamics is represented by

$$\frac{dMAP}{dt} = -\eta \frac{V_d N_e}{120 V_m} MAP + \varphi \frac{RT_{amb} C_D \pi r^2 P_{amb}}{V_m \sqrt{2RT_{amb}}} \phi_{TPS} \tag{5.2}$$

with filling dynamics time constant around 60 ms. The dynamics of the throttle response is approximated by

$$\frac{d\phi_{TPS}}{dt} = -\frac{k_{ETC}}{b_{ETC}} \phi_{TPS} + \frac{c_{ETC}}{b_{ETC}} I_{ETC} \tag{5.3}$$

where η, V_d, V_m, and N_e are volumetric efficiency of the intake process, engine displacement, intake manifold volume, and engine speed, respectively; R, T_{amb}, P_{amb}, and C_D are gas constant, ambient temperature, ambient pressure, and valve discharge constant, respectively; and ϕ_{TPS}, k_{ETC}, b_{ETC}, and c_{ETC} are engine throttle position, spring stiffness of the throttle plate, damping coefficient of the throttle plate, and throttle motor torque constant, respectively. The throttle time constant is around 50 ms. Equations (5.2) and (5.3) can be combined, discretized, and represented by the following discrete state space model:

$$x(k + 1) = Ax(k) + Bu(k)$$
$$y(k) = Cx(k) + Du(k) \tag{5.4}$$

where

$$u = I_{ETC}; \quad x = \begin{bmatrix} x_1 \\ x_2 \end{bmatrix} = \begin{bmatrix} MAP \\ \phi_{TPS} \end{bmatrix}; \quad y = MAP \tag{5.5}$$

are the system input, state, and output, respectively. The system matrices are

$$A = \begin{bmatrix} 1 - \dfrac{\eta(k)V_d N_e}{120 V_m} \Delta t & \dfrac{\varphi(k)R_a T_a C_D \pi r^2 P_a}{V_m \sqrt{2R_a T_a}} \Delta t \\ 0 & 1 - \dfrac{k_{ETC}}{b_{ETC}} \Delta t \end{bmatrix}, \quad B = \begin{bmatrix} 0 \\ \dfrac{k_{ETC}}{b_{ETC}} \Delta t \end{bmatrix}$$

$$C = \begin{bmatrix} 1 & 0 \end{bmatrix}, \qquad\qquad\qquad D = 0 \tag{5.6}$$

where Δt is the sample period. State space model (5.4) is linear time variant since the volumetric efficiency η and multiplier φ in Equations (5.2) and (5.6) are functions of the engine operating condition. Moreover, the sampling period ΔT in (5.6) equals 1 ms, and sample time index k is the same as that in Equation (5.1).

5.2.4.5 LQ Tracking Control Design

Based on the control-oriented engine MAP model, a finite horizon LQ optimal tracking controller was designed to follow the reference $z(k)$. More specifically, the control objective is to minimize the tracking error $e(k)$ defined in (5.7) with the feasible control effort I_{ETC}. The tracking error $e(k)$ is defined as

$$e(k) = y(k) - z(k) = Cx(k) - z(k) \tag{5.7}$$

and the constraint on I_{ETC} is $-5A < I_{ETC} < 5A$. The cost function of the LQ optimal controller is defined as

$$J = \frac{1}{2}[Cx(k_f) - z(k_f)]^T F[Cx(k_f) - z(k_f)] \tag{5.8}$$

$$+ \frac{1}{2} \sum_{k=k_i}^{k=k_f-1} \left\{ [Cx(k) - z(k)]^T Q[Cx(k) - z(k)] + u^T(k)Ru(k) \right\}$$

where F and Q are positive semidefinite and R is positive definite. For this design, F and Q are constant matrices defined in (5.9), and R is a function of sample index and tuned to minimize the tracking error with feasible throttle control effort (see Figure 5.13).

$$F = 10^{-8}, \quad Q = 4 \times 10^{-7}, \quad R = R(k) \tag{5.9}$$

Based on the cost function, the corresponding Hamiltonian is as follows:

$$H = \frac{1}{2}[Cx(k) - z(k)]^T Q[Cx(k) - z(k)] + \frac{1}{2}u^T(k)Ru(k) \tag{5.10}$$

$$+ p^T(k+1)[Ax(k) + Bu(k)]$$

According to [28], the necessary conditions for the extremum in terms of the Hamiltonian are represented as

$$\frac{\partial H}{\partial p^*(k+1)} = x^*(k+1) \Rightarrow x^*(k+1) = Ax^*(k) + Bu^*(k) \tag{5.11}$$

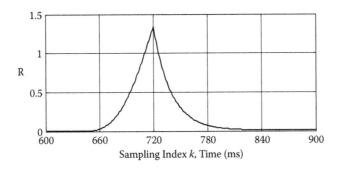

FIGURE 5.13
Adjustment of weighting matrix R.

$$\frac{\partial H}{\partial x^*(k)} = p^*(k) \Rightarrow p^*(k) = A^T p^*(k+1) + C^T QCx^*(k) - C^T Qz(k) \tag{5.12}$$

$$\frac{\partial H}{\partial u^*(k)} = 0 \Rightarrow 0 = B^T p^*(k+1) + Ru^*(k) \tag{5.13}$$

Note that the * denotes the optimal trajectories of the corresponding vectors. The augmented system of (5.11) and (5.12) becomes

$$\begin{bmatrix} x^*(k+1) \\ p^*(k) \end{bmatrix} = \begin{bmatrix} A & -BR^{-1}B^T \\ C^T QC & A^T \end{bmatrix} \begin{bmatrix} x^*(k) \\ p^*(k+1) \end{bmatrix} + \begin{bmatrix} 0 \\ -C^T Q \end{bmatrix} z(k) \tag{5.14}$$

Based on Equation (5.13), the optimal control is in the form of

$$u^*(k) = -R^{-1}B^T[P(k)x^*(k) - g(k)] \tag{5.15}$$

Matrix $P(k)$ can be computed by solving the difference Riccati equation backwards,

$$P(k) = A^T P(K+1)[I + EP(K+1)]^{-1} A + C^T QC \tag{5.16}$$

with the terminal condition

$$P(k_f) = C^T FC \tag{5.17}$$

and vector $g(k)$ can be computed by solving the vector difference equation

$$g(k) = A^T\{I - [P^{-1}(k+1) + E]^{-1}E\}g(k+1) + C^T Qz(k) \tag{5.18}$$

with the terminal condition

$$g(k_f) = C^T Fz(k_f) \tag{5.19}$$

The optimal control in Equation (5.15) can be written into the following form:

$$u^*(k) = -L_{FB}(k)x^*(k) + L_{FF}(k)g(k+1) \tag{5.20}$$

where the feedforward gain L_{FF} is computed by

$$L_{FF}(k) = [R + B^T P(k+1)B]^{-1}B^T \tag{5.21}$$

and the feedback gain L_{FB} is computed by

$$L_{FB}(k) = [R + B^T P(k+1)B]^{-1}B^T P(k+1)A \tag{5.22}$$

Note that in Equation (5.20) the state x^* used in the feedback control is computed exactly from the closed-loop system model defined below:

$$x^*(k+1) = [A - BL_{FB}(k)]x^*(k) + BL_{FF}(k)g(k+1) \tag{5.23}$$

However, when the control is implemented into the CIL simulation environment or the actual engine control system, the feedback states are replaced by the actual signals

(MAP and φ_{TPS}) measured by the onboard engine sensors. In these cases the LQ controller is represented by the online form as

$$u(k) = -L_{FB}(k)x(k) + L_{FF}(k)g(k+1) \tag{5.24}$$

where x represents the sampled states. Note that both of the states, MAP and φ_{TPS}, can be measured in the HIL simulator or in the engine system.

5.2.4.6 CIL Simulation Results and Discussion

The developed LQ optimal MAP tracking control was implemented into the Opal-RT prototype engine controller and validated through the CIL engine simulations. Figure 5.14 shows the architecture of the HIL simulation environment.

The simulated control input I_{ETC}, the system states MAP and φ_{TPS}, and λ are plotted in Figure 5.15. For comparison purposes, the simulated responses of these variables with a step I_{ETC} control are also shown in Figure 5.15, in which I_{ETC} is set to the target level before the adjustment of Π_{lift} (happens at 720 ms), and as a result, the engine throttle is gradually opened to the wide-open throttle (WOT) position and the MAP is increased before the valve lift switch. The increased MAP ensures enough fresh charge to each cylinder when the valve lift switches to the low lift. However, the step I_{ETC} control leads to a rapid

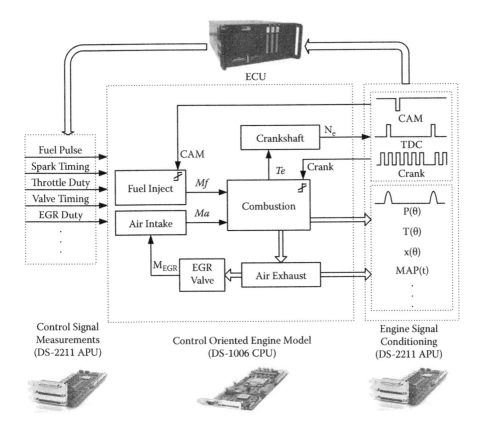

FIGURE 5.14
CIL simulation environment.

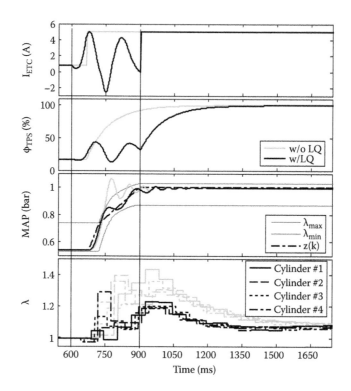

FIGURE 5.15
Engine performances of the optimal MAP tracking control.

increment of the engine MAP or excessive fresh air charge, leading to an extremely lean air-to-fuel ratio λ in the following engine cycles.

Using the proposed LQ MAP tracking control strategy, throttle current I_{ETC} is regulated in a nonmonotonic increasing pattern. Note that to maintain I_{ETC} in the feasible range ($-5A < I_{ETC} < 5A$), the weighting matrix R in the cost function (5.8) is adjusted as illustrated in Figure 5.13. The similar pattern can also be found for φ_{TPS} with a small phase lag. As a result, the engine MAP tracks the reference $z(k)$ after the intake valve lift Π_{lift} switches to the low lift, and λ of each cylinder is successfully maintained within the desired range. Therefore, with the help of the LQ optimal tracking control, the in-cylinder air-to-fuel ratio is maintained within the desired range, leading to stable combustion.

Slight oscillations in the MAP responses are found with both control approaches, which are due to the flow dynamics of the engine air handling system and the engine MAP modeling error. It is almost impossible to eliminate them. Moreover, the MAP oscillation associated with the LQ optimal tracking control is within the desired MAP range.

5.3 Control System Calibration and Integration

Control system integration involves the process of controller calibration to tune the control to provide robust stability and desired performance and to meet the fuel economy through dynamometer and vehicle tests. The dynamometer test is often considered an

HIL simulation since the dynamometer is often capable of simulating a driving cycle of an engine or vehicle to estimate the actual vehicle or engine fuel economy and emissions. After the dynamometer validation tests, vehicle calibration is aimed for for transient operation of the powertrain and engine for good drivability.

The engine calibration process consists of two major tasks: (1) fulfilling the engine feedforward control maps that are used in engine controls to meet the steady-state operation of the engine and vehicle and (2) tuning the engine control parameters as a function of engine and powertrain operational conditions to optimize the powertrain and vehicle transient operational performance with the best fuel economy possible and satisfactory emissions.

The traditional engine calibration process, shown in Figure 5.16, involves an iterative process of controller tuning and performance validation through experimental dynamometer and vehicle tests, which could be very time-consuming with high cost. On the other hand, with the demand of further improvement of vehicle fuel economy and tightened emission requirements, new engine technologies, such as VVT, EGR, variable geometry turbocharger and wastegate, and so on, have been adopted, which increases the number of engine control parameters significantly and also makes the conventional controller calibration process not feasible. Furthermore, the high competitiveness of the automotive industry leads to reduced time for product development, and as a result, the time available for controller calibration is also reduced significantly, which makes the conventional manual controller tuning and calibration process impossible. The industry is in the process of replacing the traditional calibration process with a so-called model-based or mathematically assisted calibration process, shown in Figure 5.17, where controller calibration is mainly based upon the well-calibrated dynamic engine model.

One may say that the model-based controller calibration [10] is basically replaced by the dynamic model calibration. However, the main advantage is that after the dynamic model is calibrated, generating controller calibrations is a matter of simulations instead of dynamometer tests, which reduces calibration duration and cost. To improve the dynamic model

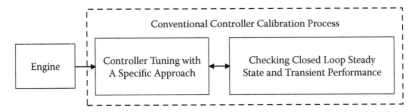

FIGURE 5.16
Conventional controller calibration process.

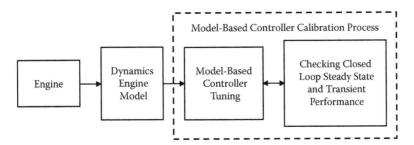

FIGURE 5.17
Model-based controller calibration process.

calibration duration, the design of experiments (DOE) [11] is often used to reduce the associated dynamic model calibration duration [29], that is, to reduce the number of dynamometer tests required to calibrate the dynamic engine using the DOE optimization technique.

References

1. S. Pace and G. Zhu, Transient Air-to-Fuel Ratio Control of an SI Engine Using Linear Quadratic Tracking, *Journal of Dynamic System, Measurement, and Control*, 136(2): 021008, 2014. DOI: 10.1115/1.4025858.
2. S. Pace and G. Zhu, Sliding Mode Control of Both Air-to-Fuel and Fuel Ratios for a Dual-Fuel Internal Combustion Engine, *Journal of Dynamic Systems, Measurement, and Control*, 134(3): 031012, 2012, DOI: 10.1115/1.4005513.
3. A. White, G. Zhu, and J. Choi, Hardware-in-the-Loop Simulation of Robust Gain-Scheduling Control of Port-Fuel-Injection Process, *IEEE Transactions on Control System Technology*, 19(2), 1433–1443, 2011. DOI: 10.1109/TCST.2010.2095420.
4. G. Zhu, I. Haskara, and J. Winkelman, Closed Loop Ignition Timing Control Using Ionization Current Feedback, *IEEE Transactions on Control System Technology*, 15(3), 2007.
5. G. Zhu, I. Haskara, and J. Winkelman, Stochastic Limit Control and Its Application to Spark Limit Control Using Ionization Feedback, presented at Proceedings of the American Control Conference, Portland, OR, June 2005.
6. C.F. Daniels, G. Zhu, and J. Winkelman, Inaudible Knock and Partial-Burn Detection Using In-Cylinder Ionization Signal, SAE Technical Paper 2003-01-3149, 2003.
7. Z. Ren and G. Zhu, Modeling and Control of an Electrical Variable Valve Timing Actuator, *Journal of Dynamic Systems, Measurement, and Control*, 136(2), 021008, 2014. DOI: 10.1115/1.4025914.
8. A. White, Z. Ren, G. Zhu, and J. Choi, Mixed H_∞ and H_2 LPV Control of an IC Engine Hydraulic Cam Phase System, *IEEE Transactions on Control System Technology*, 21(1), 229–238, 2013. DOI: 10.1109/TCST.2011.2177464.
9. I. Haskara, G. Zhu, and J. Winkelman, Multivariable EGR/Spark Timing Control for IC Engines via Extremum Seeking, presented at Proceedings of the American Control Conference, Minneapolis, MN, June 2006.
10. M. Guerrier and P. Cawsey, The Development of Model Based Methodologies for Gasoline IC Engine Calibration, SAE Technical Paper 2004-01-1466, 2014.
11. S. Jiang, D. Nutter, and A. Gullitti, Implementation of Model-Based Calibration for a Gasoline Engine, SAE Technical Paper 2012-01-0722, 2012.
12. F. Zhao, T. Asmus, D. Assanis, J.E. Dec, J.A. Eng, and P.M. Najt, *Homogeneous Charge Compression Ignition (HCCI) Engines Key Research and Development Issues*, Warrendale, PA: Society of Automotive Engineers, 2003.
13. R.M. Wagner, K.D. Edwards, et al., Hybrid SI-HCCI Combustion Modes for Low Emissions in Stationary Power Applications, presented at 3rd Annual Advanced Stationary Reciprocating Engines Meeting, Argonne, IL, June 28–30, 2006.
14. N.J. Killingsworth, S.M. Aceves, et al., HCCI Engine Combustion-Timing Control: Optimizing Gains and Fuel Consumption via Extremum Seeking, *IEEE Transactions on Control Systems Technology*, 17(6), 1350–1361, 2009.
15. C.J. Chiang and A.G. Stefanopoulou, Stability Analysis in Homogeneous Charge Compression Ignition (HCCI) Engines with High Dilution, *IEEE Transactions on Control System Technology*, 15(2), 2007.
16. X. Yang and G. Zhu, A Two-Zone Control Oriented SI-HCCI Hybrid Combustion Model for the HIL Engine Simulation, presented at Proceedings of American Control Conference, San Francisco, CA, 2011.

17. S.C. Kong and R.D. Reitz, Application of Detailed Chemistry and CFD for Predicting Direct Injection HCCI Engine Combustion and Emission, *Proceedings of the Combustion Institute*, 29: 663–669, 2002.
18. X. Yang, G. Zhu, and Z. Sun, A Control Oriented SI and HCCI Hybrid Combustion Model for Internal Combustion Engines, presented at Proceedings of ASME Dynamic Systems and Control Conference, Cambridge, MA, 2010.
19. N. Kalian, H. Zhao, and J. Qiao, Investigation of Transition between Spark Ignition and Controlled Auto-Ignition Combustion in a V6 Direct-Injection Engine with Cam Profile Switching, *Journal of Automobile Engineering*, 202, 2008.
20. N. Ravi, M.J. Roelle, et al., Model-Based Control of HCCI Engines Using Exhaust Recompression, *IEEE Transactions on Control Systems Technology*, 18(6), 1289–1302, 2010.
21. J. Kang, C. Chang, and T. Kuo, Sufficient Condition on Valve Timing for Robust Load Transients in HCCI Engines, SAE Technical Paper 2010-01-1243, 2010.
22. G.M. Shaver, Physics Based Modeling and Control of Residual-Affected HCCI Engines Using Variable Valve Actuation, PhD thesis, Stanford University, September 2005.
23. M.J. Roelle, G.M. Shaver, and J.C. Gerdes, Tackling the Transition: A Multi-Mode Combustion Model of SI and HCCI for Mode Transition Control, presented at Proceedings of IMECE International Mechanical Engineering Conference and Exposition, Anaheim, CA, November 13–19, 2004.
24. Y. Zhang, H. Xie, et al., Study of SI-HCCI-SI Transition on a Port Fuel Injection Engine Equipped with 4VVAS, SAE Technical Paper, SAE 2007-01-0199, 2007.
25. H. Wu, M. Kraft, et al., Experimental Investigation of a Control Method for SI-HCCI-SI Transition in a Multi-Cylinder Gasoline Engine, SAE Technical Paper, SAE 2010-01-1245, 2010.
26. X. Yang and G. Zhu, A Mixed Mean-Value and Crank-Based Model of a Dual-Stage Turbocharged SI Engine for Hardware-in-the-Loop Simulation, presented at Proceedings of the American Control Conference, Baltimore, MD, 2010.
27. X. Yang and G. Zhu, SI and HCCI Combustion Mode Transition Control of a Multi-Cylinder HCCI Capable SI Engine via Iterative Learning, presented at Proceedings of the 4th Annual Dynamic Systems and Control Conference, Arlington, VA, October 31–November 2, 2011.
28. D. Naidu, *Optimal Control Systems*, Boca Raton, FL: CRC Press, 2003, pp. 232–239.
29. M.-S. Vogels, J. Birnstingl, and T. Combe, Controller Calibration Using a Global Dynamic Engine Model, presented at VortragVogels Workshop, Vienna, Austria, 2011.

Index